PhoneGap 4 Mobile Application Development Cookbook

Build real-world, hybrid mobile applications using the robust PhoneGap development platform

Zainul Setyo Pamungkas

Matt Gifford

[PACKT] PUBLISHING open source community experience distilled

BIRMINGHAM - MUMBAI

PhoneGap 4 Mobile Application Development Cookbook

First published: October 2012

Second edition: October 2015

Production reference: 1261015

Published by Packt Publishing Ltd.
Livery Place
35 Livery Street
Birmingham B3 2PB, UK.

ISBN 978-1-78328-794-9

www.packtpub.com

Credits

Authors
Zainul Setyo Pamungkas
Matt Gifford

Reviewer
Mrinal Dhar

Commissioning Editor
Kevin Colaco

Acquisition Editor
Kevin Colaco

Content Development Editor
Siddhesh Salvi

Technical Editor
Tanmayee Patil

Copy Editors
Puja Lalwani
Kausambhi Majumdar
Vikrant Phadke

Project Coordinator
Nidhi Joshi

Proofreader
Safis Editing

Indexer
Rekha Nair

Production Coordinator
Manu Joseph

Cover Work
Manu Joseph

About the Authors

Zainul Setyo Pamungkas is 24-year-old full-stack developer and technology enthusiast who likes to learn and explore new things. He has been working as a web developer for over 5 years using Java and PHP. In the past 2 years, he has been interested in mobile application development. He started writing Android applications using Java and switched to PhoneGap/Cordova to create multiplatform mobile applications. He writes about development at `http://justmyfreak.com`.

Matt Gifford is an RIA developer from Cambridge, England, who specializes in ColdFusion web application and mobile development. With over 10 years of industry experience across various sectors, he is the owner of Monkeh Works. (`www.monkehworks.com`).

A regular presenter at national and international conferences, he also contributes articles and tutorials in leading international industry magazines, as well as publications on his blog (`www.mattgifford.co.uk`).

As an Adobe Community Professional for ColdFusion, Matt is an advocate of community resources and industry-wide knowledge sharing, with a focus on encouraging the next generation of industry professionals.

He is the author of *Object-Oriented Programming in ColdFusion, Packt Publishing*, and numerous open source applications, including the popular *monkehTweets* Twitter API wrapper.

First and foremost, my thanks go to all the talented PhoneGap developers for their innovative and inspiring project. Without you, this book would have been a ream of blank pages.

About the Reviewer

Mrinal Dhar is a web developer/designer, though his interests range from UX/UI design to network security and operating systems development. He created his first website when he was 9 years old, and has since worked on numerous web applications involving various languages and frameworks. He has used PhoneGap for several projects and likes the simplicity it offers to new and experienced mobile app developers. Mrinal likes working with new tech and always seeks challenging work that can push him to become better. Apart from programming, he likes photography and reading books.

He is a computer science undergraduate at IIIT Hyderabad and is expected to graduate in 2017. Mrinal intends to pursue an MS in computational linguistics once he's done with his bachelor's course.

I'd like to take this opportunity to thank my parents for everything they've done for me, which includes making me capable enough to write this here in the reviewers section of a technical book.

www.PacktPub.com

Support files, eBooks, discount offers, and more

For support files and downloads related to your book, please visit www.PacktPub.com.

Did you know that Packt offers eBook versions of every book published, with PDF and ePub files available? You can upgrade to the eBook version at www.PacktPub.com and as a print book customer, you are entitled to a discount on the eBook copy. Get in touch with us at service@packtpub.com for more details.

At www.PacktPub.com, you can also read a collection of free technical articles, sign up for a range of free newsletters and receive exclusive discounts and offers on Packt books and eBooks.

PACKTLiB™

https://www2.packtpub.com/books/subscription/packtlib

Do you need instant solutions to your IT questions? PacktLib is Packt's online digital book library. Here, you can search, access, and read Packt's entire library of books.

Why Subscribe?

- Fully searchable across every book published by Packt
- Copy and paste, print, and bookmark content
- On demand and accessible via a web browser

Free Access for Packt account holders

If you have an account with Packt at www.PacktPub.com, you can use this to access PacktLib today and view 9 entirely free books. Simply use your login credentials for immediate access.

Table of Contents

Preface

This PhoneGap 4 cookbook is a practical guide to developing hybrid applications using PhoneGap or Cordova. The sample applications here are based on real-world use cases and are shown step by step. This book covers the standard workflow using the Cordova command-line interface, and the extension of hybrid applications using various Cordova plugins. It also covers the development of hybrid applications using the Ionic framework.

What this book covers

Chapter 1, Welcome to PhoneGap 3, focuses on the new language enhancements and command-line features in PhoneGap 3.

Chapter 2, Movement and Location – Using the Accelerometer and Geolocation Sensors, is where we use built-in geolocation and accelerometer sensors. We create a mapping tool using Google Maps to display the phone's location, as well as plotting markers in the surrounding area.

Chapter 3, Filesystems, Storage, and Local Databases, looks into how to access, read, and write to and from the local file storage; list the directory contents to browse the storage; and also how to manage a local SQLite database.

Chapter 4, Working with Audio, Images, and Video, covers working with audio, images, and video, including recording/capturing and playback through local media and remote files.

Chapter 5, Working with Your Contacts List, illustrates how to manage, edit, and deal with contact information on your device.

Chapter 6, Hooking into Native Events, this chapter tells you how to override and manage native events using the PhoneGap library.

Chapter 7, Working with XUI, explains the available methods in the XUI JavaScript library.

Chapter 8, Working with the Ionic Framework, covers playing with the Ionic framework.

Chapter 9, Ionic Framework Development, look into the Ionic framework for layout and style application.

Chapter 10, User Interface Development, shows you the available mobile framework layouts, including jQuery mobile.

Chapter 11, Extending PhoneGap with Plugins, looks into creating a PhoneGap plugin across the main device formats and implementing it in the code.

Chapter 12, Development Tools and Testing, covers the setting up of your development environment and the ways to test your application.

What you need for this book

Node.js is required throughout the book. PhoneGap and the Cordova command-line interface utilize Node.js and NPM. Android Development Tool (ADT) is required if you want to build and emulate PhoneGap applications on Android. To be able to develop for the iOS platform, Xcode is needed. Unlike ADT, which can be run on major operating systems (Windows, Linux, and OS X), Xcode can only be installed on Mac OS X.

Who this book is for

If you are a developer who wants to get started with mobile application development using PhoneGap, then this book is for you. Previous experience of command-line interfaces (the terminal or Command Prompt) will help, but it is not mandatory. A basic understanding of web technologies such as HTML, CSS, and JavaScript is a must.

Sections

In this book, you will find several headings that appear frequently (*Getting ready*, *How to do it...*, *How it works...*, *There's more...*, and *See also*).

To give clear instructions on how to complete a recipe, we use these sections as follows.

Getting ready

This section tells you what to expect in the recipe, and describes how to set up any software or any preliminary settings required for the recipe.

How to do it...

This section contains the steps required to follow the recipe.

How it works...

This section usually consists of a detailed explanation of what happened in the previous section.

There's more...

This section consists of additional information about the recipe in order to make the reader more knowledgeable about the recipe.

See also

This section provides helpful links to other useful information for the recipe.

Conventions

In this book, you will find a number of text styles that distinguish between different kinds of information. Here are some examples of these styles and an explanation of their meaning.

Code words in text, database table names, folder names, filenames, file extensions, pathnames, dummy URLs, user input, and Twitter handles are shown as follows: "We can include other contexts through the use of the `include` directive."

A block of code is set as follows:

```
"installed_plugins": {
    "org.apache.cordova.network-information": {
        "PACKAGE_NAME": "com.myapp.hello"
    },
    "org.apache.cordova.battery-status": {
        "PACKAGE_NAME": "com.myapp.hello"
    }
},
"dependent_plugins": {}
```

When we wish to draw your attention to a particular part of a code block, the relevant lines or items are set in bold:

```
<script type="text/javascript" src="cordova.js"></script>
<script type="text/javascript" src="rgb.js"></script>
<script type="text/javascript">
```

Any command-line input or output is written as follows:

```
sudo npm install -g phonegap
```

New terms and **important words** are shown in bold. Words that you see on the screen, for example, in menus or dialog boxes, appear in the text like this: "If the user selects **Yes**, they can continue to press the button and see the notification window."

> Warnings or important notes appear in a box like this.

> Tips and tricks appear like this.

Reader feedback

Feedback from our readers is always welcome. Let us know what you think about this book—what you liked or disliked. Reader feedback is important for us as it helps us develop titles that you will really get the most out of.

To send us general feedback, simply e-mail `feedback@packtpub.com`, and mention the book's title in the subject of your message.

If there is a topic that you have expertise in and you are interested in either writing or contributing to a book, see our author guide at `www.packtpub.com/authors`.

Customer support

Now that you are the proud owner of a Packt book, we have a number of things to help you to get the most from your purchase.

Downloading the example code

You can download the example code files from your account at `http://www.packtpub.com` for all the Packt Publishing books you have purchased. If you purchased this book elsewhere, you can visit `http://www.packtpub.com/support` and register to have the files e-mailed directly to you.

Errata

Although we have taken every care to ensure the accuracy of our content, mistakes do happen. If you find a mistake in one of our books—maybe a mistake in the text or the code—we would be grateful if you could report this to us. By doing so, you can save other readers from frustration and help us improve subsequent versions of this book. If you find any errata, please report them by visiting `http://www.packtpub.com/submit-errata`, selecting your book, clicking on the **Errata Submission Form** link, and entering the details of your errata. Once your errata are verified, your submission will be accepted and the errata will be uploaded to our website or added to any list of existing errata under the Errata section of that title.

To view the previously submitted errata, go to `https://www.packtpub.com/books/content/support` and enter the name of the book in the search field. The required information will appear under the **Errata** section.

Piracy

Piracy of copyrighted material on the Internet is an ongoing problem across all media. At Packt, we take the protection of our copyright and licenses very seriously. If you come across any illegal copies of our works in any form on the Internet, please provide us with the location address or website name immediately so that we can pursue a remedy.

Please contact us at `copyright@packtpub.com` with a link to the suspected pirated material.

We appreciate your help in protecting our authors and our ability to bring you valuable content.

Questions

If you have a problem with any aspect of this book, you can contact us at `questions@packtpub.com`, and we will do our best to address the problem.

1

Welcome to PhoneGap 3

In this chapter, we will cover the following recipes:

- Installing PhoneGap 3
- Creating a new project
- Using the command line
- Installing API plugins

Introduction

This chapter explains the basic information about PhoneGap and how to get started with using PhoneGap. PhoneGap 3 is a big release in PhoneGap's history so far. In the older version, we had to download PhoneGap manually every time there was a new release. The pain is now over. With PhoneGap **command-line interface** (**CLI**), which was released along with PhoneGap 3, we are able to install PhoneGap directly from the command line.

PhoneGap 3 has improved the workflow for building cross-platform hybrid mobile applications. Thanks to NodeJS, creating a new project, adding a device platform, building an application, and running the application can now be performed from the command line. We don't need to open our project using each IDE, which can save us a lot of time.

Being a hybrid application means PhoneGap can give access to native functionality using web technology. We can add plugins to let our application get native capabilities. Adding plugins is easy with PhoneGap 3. Unlike older versions of PhoneGap, where we added plugins manually to each project, we can now use the CLI to add plugins to our project.

Installing PhoneGap 3

Installing PhoneGap is as easy as installing the **node package manager** (**NPM**) package. PhoneGap CLI uses NodeJS to power its command-line tool. NodeJS is a cross-platform runtime environment that uses the Google V8 JavaScript engine to execute code. NodeJS applications, including PhoneGap CLI, are written in JavaScript.

Getting ready

Before installing PhoneGap, you will need to ensure that you have all the required elements, as follows:

- A PC or Mac running Windows, OS X, or Linux. Note that you can build and run an iOS application on OS X only.
- A text editor, preferably with syntax highlighting, such as Notepad++ or Sublime Text.

How to do it...

As mentioned earlier, PhoneGap CLI is an NPM package, so we can easily install it using NPM. To install PhoneGap CLI, follow these steps:

1. First, we need to download and install NodeJS from `http://nodejs.org/` for our operating system. The installation process may be different for different operating systems. To check whether it's installed or not, you can open the terminal or Command Prompt (for Windows). Run `node -v` or `npm -v`. If you see the version number, as shown in the following screenshot, it means that you have NodeJS installed on your machine:

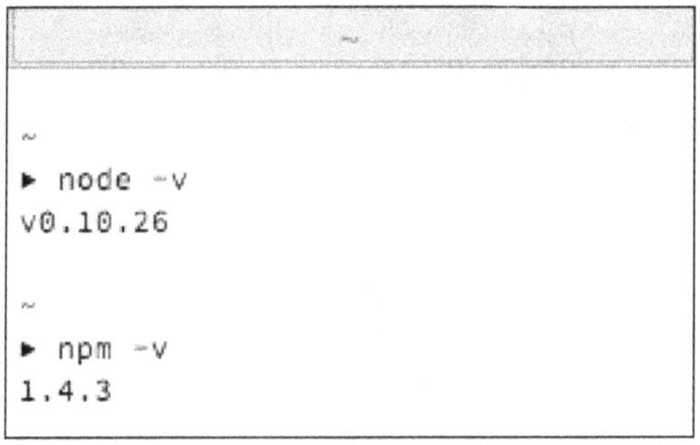

Checking the NodeJS and npm version

2. Then download and install Git client from `http://git-scm.com/` if you don't have one already. PhoneGap CLI uses Git behind the scenes to download some assets during project creation.

3. Install the `phonegap` module using `npm`. The `phonegap` module will automatically be downloaded and installed by running the following commands:
   - On Linux and OS X:

     sudo npm install -g phonegap
   - On Windows:

     npm install -g phonegap
4. Run the `phonegap` command on the terminal. You will see a help message, as follows:

```
▸ phonegap

Usage: phonegap [options] [commands]

Description:

  PhoneGap command-line tool.

Commands:

  create <path>          create a phonegap project
  serve                  serve a phonegap project
  build <platform>       build a specific platform
  install <platform>     install a specific platform
  run <platform>         build and install a specific platform
  local [command]        development on local system
  remote [command]       development in cloud with phonegap/build
  platform [command]     update a platform version
  plugin [command]       add, remove, and list plugins
  help [command]         output usage information
  version                output version number
```

Running the phonegap command to get a help message

How it works...

PhoneGap CLI is an NPM module, which is why we have to install NodeJS first. The NPM registry is located at `https://www.npmjs.org/`, and the PhoneGap CLI package is located at `https://www.npmjs.org/package/phonegap`.

The `npm install` command is a command used to install a new NPM module, and `phonegap` is the name of the module. The `npm` will search for a module named `phonegap` in the registry at `https://www.npmjs.org/`. Then, it will download the `phonegap` package along with its dependencies. We don't have to worry about which dependencies are used by `phonegap`; `npm` will do that for us. After the package has been downloaded successfully, `npm` will make the `phonegap` command available from the command line.

You might have noticed that we used a `-g` flag. This flag is used to install the module globally on our machine. It's necessary to make the `phonegap` module available globally so that we can run the `phonegap` command from anywhere, rather than only from a specific directory.

> NPM is like gem to Ruby and Composer to PHP if you have worked with Ruby or PHP before.

There's more...

It's valuable to know how NodeJS and NPM work because PhoneGap CLI is an NPM package. A public NPM package must be registered at `https://www.npmjs.org/`. Each NPM package is versioned using Git and must contain a `package.json` file in the repository. The PhoneGap CLI repository is located at `https://github.com/phonegap/phonegap-cli`.

The following JSON code is the `package.json` file from `https://github.com/phonegap/phonegap-cli/blob/master/package.json`:

```
{
    "name": "phonegap", // npm module name
    "description": "PhoneGap command-line interface and node.js
    library.", // module description
    "version": "3.5.0-0.21.18", // version number
    "homepage": "http://github.com/phonegap/phonegap-cli", //
    module homepage
    // repository type and url.
    "repository": {
        "type": "git",
        "url": "git://github.com/phonegap/phonegap-cli.git"
    },
    // module keywords
    "keywords": [
        "cli",
        "cordova",
        "phonegap",
        "phonegap build",
        "phonegap/build"
    ],
    // global installation is preferred
    "preferGlobal": "true",
    // main js file
    "main": "./lib/main.js",
    // binary code for the module.
    "bin": {
```

```
        "phonegap": "./bin/phonegap.js" // phonegap command will
        use ./bin/phonegap.js
    },
    // script is command for certain action. in this case running
    npm test will run jasmine-node --color spec
    "scripts": {
        "test": "jasmine-node --color spec"
    },
    "engineStrict": "true", // force to use specific node engine
    "engines": {
        "node": ">=0.10.0" // node engine is set to min of version
        0.10.0
    },
    // module dependencies
    "dependencies": {
        "colors": "0.6.0-1",
        "cordova": "3.5.0-0.2.7",
        "cordova-lib": "0.21.7",
        "connect-phonegap": "0.13.0",
        "minimist": "0.1.0",
        "phonegap-build": "0.8.4",
        "pluralize": "0.0.4",
        "prompt": "0.2.11",
        "qrcode-terminal": "0.9.4",
        "semver": "1.1.0",
        "shelljs": "0.1.4"
    },
    // in-development module dependencies.
    "devDependencies": {
        "jasmine-node": "1.14.5",
        "chdir": "0.0.x"
    },
    // module contributors
    "contributors": [
        {
            "name": "Michael Brooks",
            "email": "michael@michaelbrooks.ca",
            "url": "http://michaelbrooks.ca/"
        },
        {
            "name": "Lorin Beer",
            "email": "lorin.beer@gmail.com",
            "url": "http://www.ensufire.com/"
        },
```

```
            {
                "name": "Jesse MacFadyen",
                "url": "http://risingj.com"
            },
            {
                "name": "Ryan Stewart",
                "email": "ryan@adobe.com"
            }
        ]
    }
```

> **Downloading the example code**
>
> You can download the example code files from your account at `http://www.packtpub.com` for all the Packt Publishing books you have purchased. If you purchased this book elsewhere, you can visit `http://www.packtpub.com/support` and register to have the files e-mailed directly to you.

NPM will download and install every dependency referenced by `package.json`. You will notice the `cordova` module among the `dependencies`. NPM will download the `cordova` module too, so you can run `cordova` commands from your command line.

> If somehow the `cordova` module is not installed on your machine after installing PhoneGap CLI, you can install it manually by running `npm i -g cordova` from the terminal.
>
> Apache Cordova (`http://cordova.apache.org/`) is the software underlying PhoneGap. Previously, PhoneGap and Cordova were one software project. But after Adobe acquired Nitobi, the original developer of PhoneGap, PhoneGap code was contributed to the Apache Software Foundation as Apache Cordova.

Creating a new project

Creating a new PhoneGap project is easy thanks to PhoneGap CLI. Unlike older versions of PhoneGap, where we needed to download the project template manually, we can create new projects directly from the command line. With just a single command, we can create a new project and start writing our application.

How to do it...

To create a new PhoneGap project, open your terminal or `cmd`. Then go to the directory where you want to maintain your source code. Run the following command:

```
phonegap create hello com.myapp.hello HelloWorld
```

It may take some time to complete the process, so be patient and let PhoneGap CLI do its magic. You will see some message during the project creation process:

```
phonegap-cookbook/chapter1/sample1
▸ phonegap create hello com.myapp.hello HelloWorld
[phonegap] the options /Users/justmyfreak/Desktop/playground/phonegap-cookbook/c
hapter1/sample1/hello com.myapp.hello HelloWorld
[phonegap] created project at /Users/justmyfreak/Desktop/playground/phonegap-coo
kbook/chapter1/sample1/hello
```

Creating a new project progress

Congratulations! You have created your first PhoneGap project. Now, let's browse the project directory. You will see the directories as shown in the following screenshot:

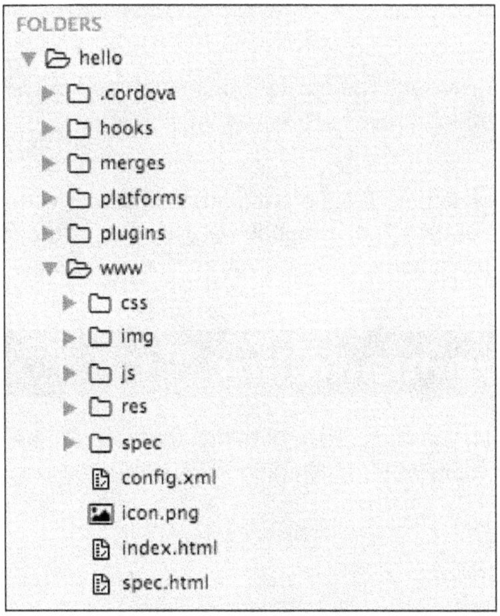

The PhoneGap project structure

Most of the directories are empty; we will discuss the use of each directory later in the next recipe. The `www/` directory is where you write code for your application. PhoneGap generated the initial starter app to work with.

How it works...

The `phonegap create` command is a command used to create a new project. The first argument, `hello`, specifies a directory for your project. Note that this directory must not exist initially; `phonegap` will create it for you.

The second argument, `com.myapp.hello`, is your application ID. The application ID is used as a unique identifier for an application. Two identical applications with different application IDs will be considered two different applications. The application ID is in reverse domain style. You can create something like `com.yourdomain.applicationname`.

The third argument, `HelloWorld`, is your application name. The application name will be used as the application's display title. This argument is optional. If you are not setting the application name, it will use the name from the first argument. If you want to change the name, you can open `config.xml` and edit the `name` element.

While we are running the `phonegap create` command, there are several things happening in the background:

- The `phonegap` creates a new PhoneGap project with the given name and ID in the newly created directory. In our case, a PhoneGap project with the name as `HelloWorld` and ID as `com.myapp.hello` will be created under the `hello` directory.
- The `phonegap` downloads the starter application and places it in `www/` so that we can run the project directly after creating it.

> You can use the `-d` option with any `phonegap` command to allow a verbose output. The `-d` option will give clear information about what is going on and the current status of the command.

Using the command line

After creating a new project, there are several things that need to be done before we are able to run the project. The workflow of PhoneGap consists of the following mandatory steps:

1. Creating new project.
2. Adding device platform.
3. Building the project.
4. Running the project.

How to do it...

The PhoneGap command consists of two environments. The first is the local command environment. The local commands execute the command on your local machine. In this case, you must have the target device SDKs configured on your machine. For example, if you want to develop an Android application, you must acquire and configure the Android SDK on your machine.

The second environment is remote. Command-line commands execute the build process remotely using the cloud-based **PhoneGap Build** service. In this case, you don't need to configure any SDK on your local machine.

The local commands

We created our first PhoneGap project in the previous recipe. The next thing to do is explore the `phonegap` commands. Follow these steps to learn about the `phonegap` commands that will be used to get your application running:

1. Change the directory to your project directory. After creating a new project, simply run `cd hello` to go to your project's directory. All further `phonegap` commands need to be run in the project's directory.
2. Before we can build and run our project, we have to add target platforms. The command used to add the platform is as follows:

   ```
   cordova platform add <target name>
   ```

3. The `<target name>` argument is your target platform. The ability to build and run a project on your machine depends on the SDK availability for your machine. On Mac, you can run the following commands to add a specific platform:

   ```
   cordova platform add ios
   cordova platform add android
   cordova platform add amazon-fireos
   cordova platform add blackberry10
   cordova platform add firefoxos
   ```

   On Windows, you can add these platforms:

   ```
   cordova platform add wp7
   cordova platform add wp8
   cordova platform add wp7
   cordova platform add windows8
   cordova platform add amazon-fireos
   cordova platform add blackberry10
   cordova platform add firefoxos
   ```

4. You may have noticed that we are using `cordova` instead of `phonegap` to add target platforms. I mentioned that `cordova` is one of the `phonegap` dependencies, so all `cordova` commands are available for the `phonegap` project.
5. Let's add the Android platform for our project by running the `cordova` platform. Add `android`. Now browse through your project using the file explorer. You will see the `android` directory inside the `platforms/` directory. The `android` directory is an Android project. You can open it on IDEs such as Eclipse or IntelliJ. The same thing happens when we add the iOS platform. We will get the `ios` directory inside `platforms` along with the Xcode project files:

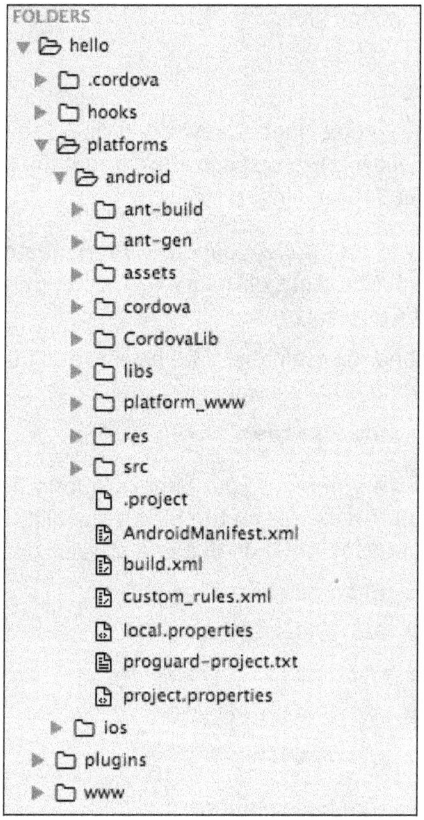

The native project inside the platforms/ directory

> When PhoneGap and Cordova release a new version, we should update our platform-specific project by running `phonegap platform update <platform>`.

6. The next step is to build the project. Building the project means compiling your application into byte code for the target device. Run the following command to build your project:

   ```
   phonegap build <platform>
   ```

7. A PhoneGap application can run on the local development server. We can save time by testing our application on the local development server instead of running on an emulator or a real device. To run an application on the local web server, run the following command:

   ```
   phonegap serve
   ```

8. The `serve` command has some options:

   ```
   --port, -p <n>        Port for web server. Default port is 3000
   --autoreload         Enable live-reload when file changes occur.
   Default is true
   --no-autoreload   Disable live-reload on file changes.
   --localtunnel         Enabling local tunnel, will expose the local
   server to the your network. Default value is false
   ```

9. For example, to serve an application on port `1337` without auto-reload, you can run the following command:

   ```
   phonegap serve -p 1337 —no-autoreload
   ```

10. Install the application platform by running this line:

    ```
    phonegap install <platform>
    ```

11. The `install` command has some options:

    ```
    —device              Force installation to device
    —emulator          Force installation to emulator
    ```

12. By default, `phonegap` will check whether you have a connected device or not. If a connected device is found, your application will be installed on that device. If no device is connected or found, `phonegap` will install it on the emulator. You don't need to open your emulator; `phonegap` will run it for you:

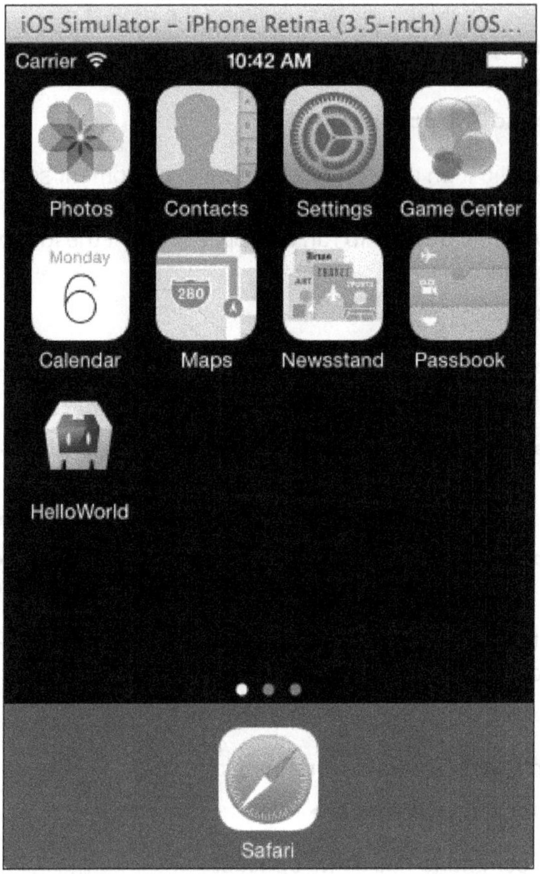

A PhoneGap starter application installed on an iOS simulator

> To be able to run an application directly on your Android device, you have to enable USB debugging on that device.

13. To run your application, you can use the following command:

```
phonegap run <platform>
```

14. The `run` command has some options:

```
–device            Force application to run on device
–emulator          Force application to run on emulator
```

15. Just as with the `install` command, `phonegap run` will check whether you have a connected device or not, by default. If a connected device is found, your application will be run on that device. Otherwise, `phonegap` will run it on an emulator:

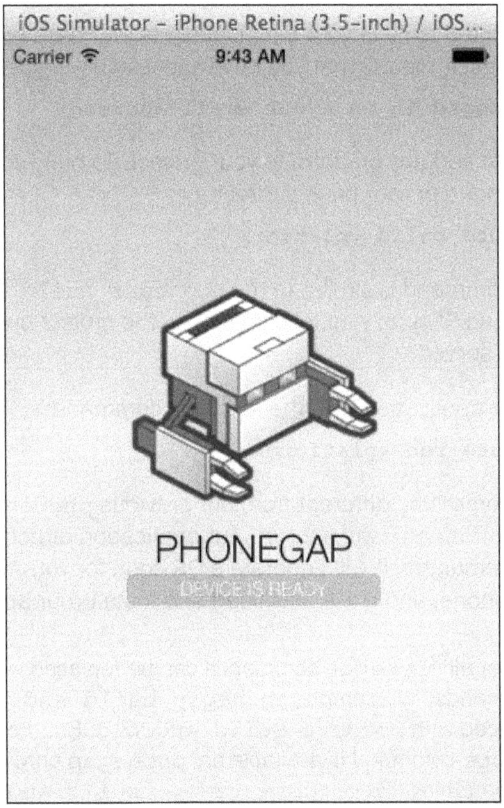

A PhoneGap starter application running on an iOS simulator

Congratulations!!! You have created your first PhoneGap application and successfully launched it.

The remote commands

The PhoneGap CLI also ships a tool for integrating our application with PhoneGap Build. PhoneGap Build allows you to build your hybrid application without setting up any SDK on your local machine. Using PhoneGap Build, you can create an iOS application from Windows and create a Windows Mobile application from OS X.

To be able to use PhoneGap Build and the phonegap remote command, you have to sign up at http://build.phonegap.com. Make sure that you don't use the GitHub single sign-on. The phonegap remote doesn't support GitHub accounts. To use the phonegap remote commands, simply follow these instructions:

1. Log in to your PhoneGap Build account. Simply open your command line and run this command:

   ```
   phonegap remote login
   ```

2. You will be prompted to fill in your e-mail address and password. If you see the following message, it means that you have successfully logged in:

   ```
   [phonegap] logged in as <your email address>
   ```

3. You don't need to add any platform to your project. To build your application, you have to run this command in your project directory:

   ```
   phonegap remote build <platform>
   ```

4. The preceding command is similar to the phonegap build <platform> command. Instead of using an SDK on your local machine, the project build takes place on the PhoneGap Build server.

5. You can run your application with the following command:

   ```
   phonegap remote run <platform>
   ```

6. You will notice something different from our previous phonegap run <platform> command. The phonegap will not run your application directly, whether on your device or on an emulator. It will generate a QR code for you. You can scan the QR code with your phone, and it will download and install your application in your device.

> Almost all phonegap commands can be replaced with cordova commands; for example, phonegap build android can be replaced with cordova build android. But the phonegap remote command is available on phonegap only. So you can't do something like cordova remote build android.
>
> If you are confused between Cordova and PhoneGap, read about the history of Cordova and PhoneGap at http://phonegap.com/2012/03/19/phonegap-cordova-and-what's-in-a-name/.

How it works...

When building a PhoneGap application, the first thing to do is create a new PhoneGap project. PhoneGap will generate the project files automatically and give a starter application as a sample. Then we add platforms that we want to work with. For a remote-based build, we don't need to specify which platforms we want to work with. PhoneGap Build will prepare our application to work with iOS, Android, and Windows Mobile.

The main difference between local-based builds and remote-based builds is where the application building process is done. For local-based builds, PhoneGap will use the platform SDK that is installed on your machine to build the application. If the SDK is not found, PhoneGap will force you to use a remote-based build.

In the case of a remote-based build, your code will be uploaded to the PhoneGap Build server. Then the PhoneGap Build server will build your application using its machine. We do not get a device-specific project, such as an Android Eclipse project or iOS Xcode project, when using remote-based builds. What we get is a distribution-ready binary. We will get `*.ipa` for an iOS application and `*.apk` for an Android application.

Installing API plugins

A plugin is a piece of add-on code that provides an interface for native components. A plugin contains native code and a JavaScript interface. Using plugins, we can access native features using JavaScript code. We can get access to a camera, a file browser, geolocation, and so on by calling the PhoneGap JavaScript API.

How to do it...

The PhoneGap CLI allows us to manage plugins easily from the command line. Adding new plugins and removing existing plugins is easy now. We don't have to download and configure a plugin for each platform that we are targeting. Prior to PhoneGap 3, plugin management was a pain.

Adding plugins

When creating a new PhoneGap project, PhoneGap doesn't include any plugins in the project. It makes our initial application clean. First, we may want to build an application without native capabilities, just like developing a web application. Then we can add plugins to extend the application.

To add a plugin to an existing project, run the following command in your project directory:

```
phonegap plugin add <source>
```

The `<source>` argument can be the path to the `plugin` directory on a local machine, the `git` repository, or the `plugin` namespace. The following commands can be used to add a plugin from the various sources mentioned before:

```
phonegap plugin add /local/path/to/plugin/
phonegap plugin add http://example.com/path/to/plugin.git
phonegap plugin add org.apache.cordova.device
```

Once a plugin has been successfully added to a project, the plugin APIs can be executed using JavaScript. Each plugin has its own way of accessing native APIs, so read the documentation for each plugin.

> You can search for an existing plugin using the `cordova plugin search <keyword>` command.

Listing plugins

After installing plugins, you can list all the installed plugins by running the following command:

```
phonegap plugin list
```

You will see a list of plugins installed, like this:

```
chapter1/sample1/hello
▸ phonegap plugin list
[phonegap] the following plugins are installed
org.apache.cordova.battery-status 0.2.11 "Battery"
org.apache.cordova.device 0.2.12 "Device"
org.apache.cordova.network-information 0.2.12 "Network Information"
```

Removing plugins

To remove the installed plugins, simply run the following command from your project directory:

```
phonegap plugin remove <id>
```

The `<id>` argument is the plugin `id` inside the plugin's `plugin.xml` file. The `<id>` argument is also the name of the `plugin` directory inside the `plugins/` folder in your project. For example, if we want to remove the `org.apache.cordova.device` plugin, we can run this command:

```
phonegap plugin remove org.apache.cordova.device
```

How it works...

When the `phonegap plugin add` command is executed, `phonegap` will copy the plugin files from the source to the project under the `plugins/` directory. Each plugin will have its own directory, with the plugin ID as the directory name. Inside each `plugin` folder, you will find the `doc/`, `src/`, `tests/`, and `www/` directories, along with other files.

The `doc/` folder contains the plugin documentation. The `src/` folder contains native code for each platform. You will see Java code for Android, Objective-C code for iOS, and so on. The `tests/` folder contains JavaScript unit tests for the JavaScript interface. The last folder is `www/`. It contains markup, styling, media, and JavaScript code that is used for presentation and to interface with native code. The main code of the PhoneGap application will be placed in the `www` folder.

After the plugin is copied to the `plugins` directory, `phonegap` will update or create a `.json` file inside `plugins/`. Each platform will have its own `.json` file. Android will have `android.json`, while iOS will have `ios.json`. These `.json` files hold the plugin configuration for each platform. The following is an example of the use of plugin configurations:

- **Adding a new permission**: Some plugins may need to add permissions to be able to work properly. Here is an example of modifying `AndroidManifest.xml` for the Android project. A new `ACCESS_NETWORK_STATE` permission is added so that we can have access to the network state:

```
"AndroidManifest.xml": {
    "parents": {
        "/*": [
            {
                "xml": "<uses-permission
android:name=\"android.permission.ACCESS_NETWORK_STATE\"
/>",
                "count": 1
            }
        ]
    }
}
```

- **Modifying platform project files**: Some plugins may need to add some configurations to each platform project. The following is an example of a configuration for modifying `res/xml/config.xml` in the Android project:

```
"res/xml/config.xml": {
    "parents": {
        "/*": [
            {
```

```
                    "xml": "<feature
name=\"NetworkStatus\"><param name=\"android-package\"
value=\"org.apache.cordova.networkinformation
.NetworkManager\" /></feature>",
                    "count": 1
                },
                {
                    "xml": "<feature name=\"Battery\">
                    <param name=\"android-package\"
                    value=\"org.apache.cordova.batterystatus
                    .BatteryListener\" /></feature>",
                    "count": 1
                }
            ]
        }
}
```

- **Declaring which plugins are used and plugin dependency**: The configuration also holds information about which installed plugins are used for each platform:

```
"installed_plugins": {
    "org.apache.cordova.network-information": {
        "PACKAGE_NAME": "com.myapp.hello"
    },
    "org.apache.cordova.battery-status": {
        "PACKAGE_NAME": "com.myapp.hello"
    }
},
"dependent_plugins": {}
```

See also

- *Chapter 2, Movement and Location – Using the Accelerometer and Geolocation Sensors*
- *Chapter 3, Filesystems, Storage, and Local Databases*
- *Chapter 4, Working with Audio, Images, and Video*
- *Chapter 5, Working with Your Contacts List*
- *Chapter 6, Hooking into Native Events*
- *Chapter 11, Extending PhoneGap with Plugins*

2

Movement and Location – Using the Accelerometer and Geolocation Sensors

In this chapter, we will cover the following recipes:

- Detecting device movement using the accelerometer
- Adjusting the accelerometer sensor update interval
- Updating a display object position through accelerometer events
- Obtaining device geolocation sensor information
- Adjusting the geolocation sensor update interval
- Retrieving map data through geolocation coordinates
- Creating a visual compass to show the device direction

Introduction

Mobile devices are incredibly powerful tools that not only allow us to make calls and send messages, but also help us navigate and identify where we are in the world, thanks to the accelerometer, geolocation, and other sensors.

This chapter will explore how we can access these sensors and make use of this exposed functionality using plugins.

Detecting device movement using the accelerometer

The accelerometer captures device motion in the *x*, *y*, and *z* axis direction. The accelerometer is a motion sensor that detects change (delta) in movement relative to the orientation of the current device.

How to do it...

We will use the accelerometer functionality from the plugins to monitor the feedback from the device:

1. First, create a new PhoneGap project named `accelerometer`. Open Terminal or Command Prompt and run the following command:

   ```
   phonegap create accelerometer com.myapp.accelerometer
   accelerometer
   ```

2. Add the device's platform. You can choose to use Android, iOS, or both:

   ```
   cordova platform add ios
   cordova platform add android
   ```

3. Add the `device-motion` plugin by running the following command:

   ```
   phonegap plugin add org.apache.cordova.device-motion
   ```

4. Open `www/index.html` and clean up unnecessary elements; so you have the following:

   ```
   <!DOCTYPE html>
   <html>
       <head>
           <meta charset="utf-8" />
           <meta name="format-detection"
           content="telephone=no" />
           <meta name="msapplication-tap-highlight"
           content="no" />
           <meta name="viewport" content="user-scalable=no,
           initial-scale=1, maximum-scale=1, minimum-scale=1,
           width=device-width, height=device-height,
           target-densitydpi=device-dpi" />
           <link rel="stylesheet" type="text/css"
           href="css/index.css" />
           <title>Hello World</title>
       </head>
       <body>
   ```

```
        <h1>Accelerometer Data</h1>
        <div id="accelerometerData">Obtaining data…</div>
        <script type="text/javascript"
        src="cordova.js"></script>
        <script type="text/javascript">

        </script>
    </body>
</html>
```

5. Below the Cordova JavaScript reference, write a new JavaScript tag block and define an event listener to ensure that the device is ready and the native code has loaded before continuing:

```
<script type="text/javascript">
      // Set the event listener to run
      // when the device is ready
      document.addEventListener('deviceready',
      onDeviceReady, false);
</script>
```

6. Now add in the `onDeviceReady` function, which will run the `getCurrentAcceleration` method when the native code has fully loaded:

```
// The device is ready so let's
// obtain the current accelerometer data
function onDeviceReady() {
     navigator.accelerometer.getCurrentAcceleration(
     onSuccess, onError);
}
```

7. Include the `onSuccess` function to handle the returned information from the accelerometer.
8. Define the `accelerometer div` element to the `accElement` variable to hold the generated accelerometer results.
9. Next, assign the returned values from the `acceleration` object as the HTML within the `accelerometer div` element for display to the user; the available properties are accessed through the `acceleration` object:

```
// Run after successful transaction
// Let's display the accelerometer data
function onSuccess(acceleration) {
    var accElement =
        document.getElementById('accelerometerData');

    accElement.innerHTML    =
        'Acceleration X: ' + acceleration.x + '<br />' +
```

```
            'Acceleration Y: ' + acceleration.y + '<br />' +
              'Acceleration Z: ' + acceleration.z + '<br />' +
              'Timestamp: '      + acceleration.timestamp;
    }
```

10. Finally, include the `onError` function to deal with any possible issues:

```
// Run if we face an error
// obtaining the accelerometer data
function onError(error) {
    // Handle any errors we may face
    alert('error');
}
```

11. Build and run the project:

```
cordova build android
cordova run android
```

12. When running the application on an emulator, the output will look something like this:

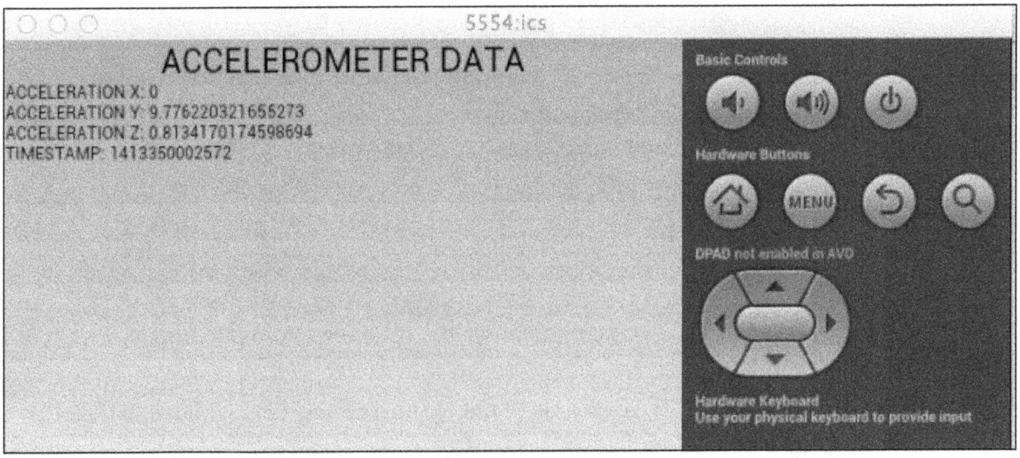

How it works...

By registering an event listener to the `deviceready` event, we are ensuring that the JavaScript code does not run before the native PhoneGap code is executed. Once ready, the application will call the `getCurrentAcceleration` method from the accelerometer API, providing two methods to handle successful transactions and errors respectively.

The onSuccess function returns the obtained acceleration information in the form of the following four properties:

- acceleration.x: This is a Number registered in meters per second squared (m/s^2) that measures the device acceleration across the *x* axis. This is the movement from left to right when the device is placed with the screen in an upright position. Positive acceleration is obtained as the device is moved to the right, whereas a negative movement is obtained when the device is moved to the left.
- acceleration.y: This is a Number registered in m/s^2 that measures the device acceleration across the *y* axis. This is the movement from bottom to top when the device is placed with the screen facing an upright position. Positive acceleration is obtained as the device is moved upwards, whereas a negative movement is obtained when the device is moved downwards.
- acceleration.z: This is a Number registered in m/s^2 that measures the device acceleration across the *z* axis. This is perpendicular from the face of the device. Positive acceleration is obtained when the device is moved to face towards the sky, whereas a negative movement is obtained when the device is pointed towards the Earth.
- acceleration.timestamp: This is a DOMTimeStamp that measures the number of milliseconds from the point of the application's initialization. This could be used to store, update and track changes over a period of time since the last accelerometer update.

The following diagram shows the **X**, **Y**, and **Z** axis in relation to the device:

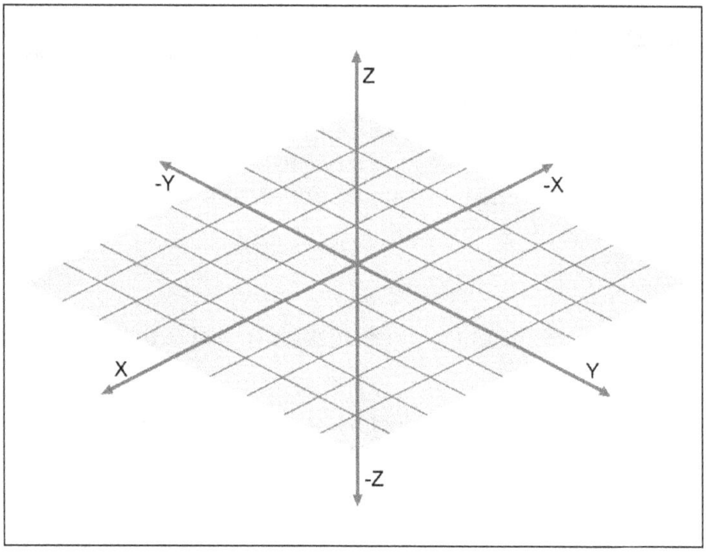

The `acceleration.x`, `acceleration.y`, and `acceleration.z` values returned from the aforementioned acceleration object include the effect of gravity, which is defined as precisely 9.81 m/s^2.

> iOS doesn't know about the concept of getting current acceleration at any given point. Device motion must be watched and the data captured at a given time interval.

There's more...

Accelerometer data obtained from devices has been used to great effect in mobile handset games that require balance control and detection of movement, including steering, control views, and tilting objects. You can check out the official Cordova documentation covering the `getCurrentAcceleration` method and obtaining accelerometer data.

See also

- You can check out the official Cordova documentation covering the `getCurrentAcceleration` method and obtaining accelerometer data at `http://plugins.cordova.io/#/package/org.apache.cordova.device-motion`.

Adjusting the accelerometer sensor update interval

The `getCurrentAcceleration` method obtains data from the accelerometer at the time it was called - a single call to obtain a single response object. In this recipe, we'll build an application that allows us to set an interval to obtain a constant update from the accelerometer to detect continual movement from the device.

How to do it...

We will provide additional parameters to a new method available through the PhoneGap API to set the update interval:

1. First, create a new PhoneGap project named `accelupdate`. Open Terminal or Command Prompt and run the following command:

   ```
   phonegap create accelupdate com.myapp.accelupdate accelupate
   ```

2. Add the devices platform. You can choose to use Android, iOS, or both:

   ```
   cordova platform add ios
   cordova platform add android
   ```

3. Add the `device-motion` plugin by running the following command:

   ```
   phonegap plugin add org.apache.cordova.device-motion
   ```

4. Open `www/index.html` and clean up unnecessary elements; so you have the following:

   ```
   <!DOCTYPE html>
   <html>
       <head>
           <meta charset="utf-8" />
           <meta name="format-detection"
           content="telephone=no" />
           <meta name="msapplication-tap-highlight"
           content="no" />
           <meta name="viewport" content="user-scalable=no,
           initial-scale=1, maximum-scale=1, minimum-scale=1,
           width=device-width, height=device-height,
           target-densitydpi=device-dpi" />
           <link rel="stylesheet" type="text/css"
           href="css/index.css" />
           <title>Hello World</title>
       </head>
       <body>
           <h1>Accelerometer Data</h1>
           <div id="accelerometerData">Obtaining data…</div>

           <script type="text/javascript"
           src="cordova.js"></script>
           <script type="text/javascript">

           </script>
       </body>
   </html>
   ```

5. Below the Cordova JavaScript reference, write a new JavaScript tag block, within which a variable called `watchID` will be declared.

6. Next, define an event listener to ensure that the device is ready and the native code has loaded before continuing:

   ```
   <script type="text/javascript">
       // The watch id variable is set as a
       // reference to the current 'watchAcceleration'
       var watchID = null;
       // Set the event listener to run
       // when the device is ready
   ```

```
        document.addEventListener('deviceready',
        onDeviceReady, false);
</script>
```

7. Now add in the `onDeviceReady` function, which will run a method called `startWatch` once the native code has fully loaded:

```
// The device is ready so let's
// start watching the acceleration
 function onDeviceReady() {
        startWatch();
}
```

8. Next, write the `startWatch` function. To do this, first, create a variable called `options` to hold the optional `frequency` parameter, set to 3,000 milliseconds (3 seconds).
9. Then, set the initial `disabled` properties of two buttons that will allow the user to start and stop the acceleration detection.
10. Next, assign the `watchAcceleration` to the previously defined `watchID` variable. This will allow you to check for a value or identify if it is still set to `null`.
11. Apart from defining the success and error function names, you are also sending the `options` variable into the method call, which contains the `frequency` value:

```
// Watch the acceleration at regular
// intervals as set by the frequency
function startWatch() {

    // Set the frequency of updates
    // from the acceleration
    var options = { frequency: 3000 };

    // Set attributes for control buttons
    document.getElementById('startBtn').disabled = true;
    document.getElementById('stopBtn').disabled = false;

    // Assign watchAcceleration to the watchID variable
    // and pass through the options array
    watchID = navigator.accelerometer.watchAcceleration(
    onSuccess, onError, options);
}
```

12. With the `startWatch` function written, you now need to provide a method to stop the detection of the acceleration. This firstly checks the value of the `watchID` variable. If this is not `null`, it will stop watching the acceleration using the `clearWatch` method, passing in the `watchID` parameter before resetting this variable back to `null`.

13. Then reference the `accelerometer div` element and set its value to a user-friendly message.
14. Next, reassign the `disabled` properties for both the control buttons to allow the user to start watching again, as follows:

```
// Stop watching the acceleration
function stopWatch() {

    if (watchID) {
        navigator.accelerometer.clearWatch(watchID);
      watchID = null;

        var element =
            document.getElementById('accelerometerData');

        element.innerHTML =
            'No longer watching your acceleration.';

        // Set attributes for control buttons
        document.getElementById('startBtn').disabled =
        false;
        document.getElementById('stopBtn').disabled = true;

    }
}
```

15. Now, create the `onSuccess` method, which will be run after a successful update response. Assign the returned values from the `acceleration` object as the HTML within the `accelerometer div` element for display to the user—the available properties are accessed through the `acceleration` object and applied to the string variable:

```
// Run after successful transaction
// Let's display the accelerometer data
function onSuccess(acceleration) {
    var element =
    document.getElementById('accelerometerData');
    element.innerHTML =
        'Acceleration X: ' + acceleration.x + '<br />' +
        'Acceleration Y: ' + acceleration.y + '<br />' +
      'Acceleration Z: ' + acceleration.z + '<br />' +
      'Timestamp: '      + acceleration.timestamp +
      '<br />';
}
```

16. You also need to supply the `onError` method to catch any possible issues with the request. Here, output a user-friendly message, setting it as the value of the `accelerometerData` `div` element:

```
// Run if we face an error
// obtaining the accelerometer data
function onError() {
    // Handle any errors we may face
    var element =
    document.getElementById('accelerometerData');
    element.innerHTML =
        'Sorry, I was unable to access the acceleration
        data.';
}
```

17. Finally, add in the two button elements, both of which will have an `onClick` attribute set to either `start` or `stop` watching the device acceleration:

```
<body>
    <h1>Accelerometer Data</h1>

    <button id="startBtn"
    onclick="startWatch()">start</button>

    <button id="stopBtn"
    onclick="stopWatch()">stop</button>

    <div id="accelerometerData">Obtaining data...</div>

</body>
```

18. The results will appear similar to the following screenshot:

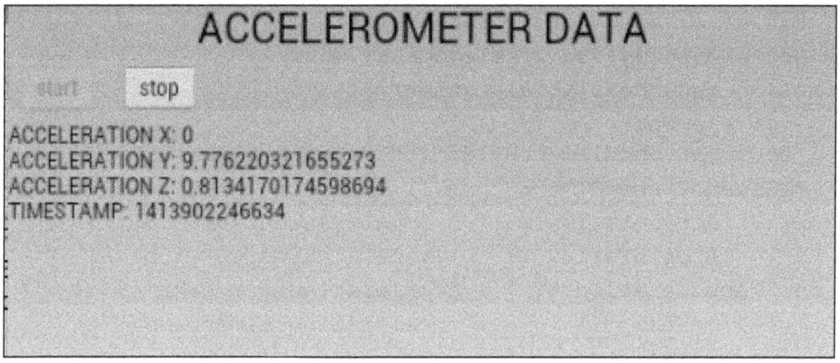

19. Stopping the acceleration watch will result something like this:

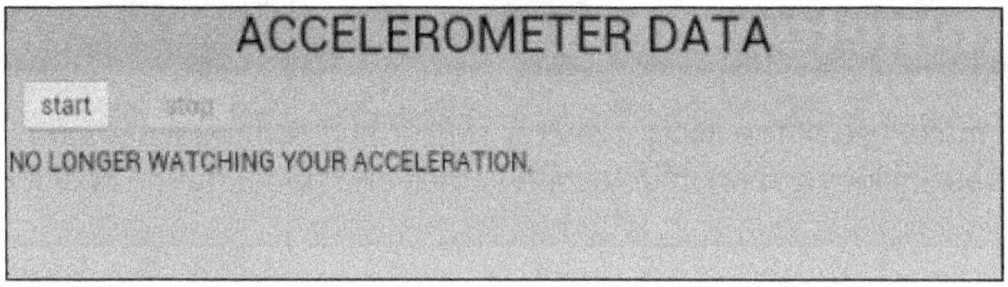

How it works...

By registering an event listener to the `deviceready` event, we ensure that the JavaScript code does not run before the native PhoneGap code is executed. Once ready, the application calls the `startWatch` function, within which the desired frequency interval for the acceleration updates is set.

The `watchAcceleration` method from the PhoneGap API retrieves the device's current acceleration data at the specified interval. If not, as the interval is passed through, it defaults to 10,000 milliseconds (10 seconds). Each time an update has been obtained, the `onSuccess` method is run to handle the data as you wish—in this case, displaying the results on the screen.

The `watchID` variable contains a reference to the `watch` interval and is used to stop the watching process by passing it into the `clearWatch` method from the PhoneGap API.

There's more...

In this example, the `frequency` value for the accelerometer update interval was set at 3,000 milliseconds (3 seconds). Consider writing a variation on this application that allows the user to manually change the interval value using a slider, or by setting the desired value into an input box.

See also...

- You can check out the official Cordova documentation covering the `watchAcceleration` method and specifying how to obtain accelerometer data at `http://plugins.cordova.io/#/package/org.apache.cordova.device-motion`.

Updating a display object position through accelerometer events

Developers can make use of the accelerometer sensor and continual updates provided by it for many purposes, including motion-detection games and updating the position of an object on the screen.

How to do it...

We will use the device's accelerometer sensor on continual update to move an element around the screen as a response to device movement:

1. First, create a new PhoneGap project named `accelobject` by running the following command:

   ```
   phonegap create accelobject com.myapp.accelobject accelobject
   ```

2. Add the device platform. You can choose to use Android, iOS, or both:

   ```
   cordova platform add ios
   cordova platform add android
   ```

3. Add the device motion plugin by running the following command:

   ```
   phonegap plugin add org.apache.cordova.device-motion
   ```

4. Open `www/index.html` and clean up unnecessary elements. Within the `body` tag create two `div` elements. Set the first with the `id` attribute equal to `dot`. This will be the element that will move around the screen of the device.
5. The second `div` element will have the `id` of `accelerometerData` and will be the container into which the returned acceleration data will be output:

   ```
   <html>
       <head>
           <meta charset="utf-8" />
           <meta name="format-detection"
           content="telephone=no" />
           <meta name="msapplication-tap-highlight"
           content="no" />
           <meta name="viewport" content="user-scalable=no,
           initial-scale=1, maximum-scale=1, minimum-scale=1,
           width=device-width, height=device-height,
           target-densitydpi=device-dpi" />
           <title>Hello World</title>
       </head>
   ```

```
    <body>
        <h1>Accelerometer Movement</h1>

        <div id="dot"></div>

        <div id="accelerometerData">Obtaining data...</div>

        <script type="text/javascript"
        src="js/index.js"></script>
        <script type="text/javascript">

        </script>
    </body>
</html>
```

6. You can now start with the custom scripting and PhoneGap implementation. Add a `script` tag block before the closing `head` tag to house the code:
7. Before diving into the core code, you need to declare some variables. Here, set a default value for `watchID` as well as the radius for the circle display object you will be moving around the screen:

```
// The watch id variable is set as a
// reference to the current `watchAcceleration`
var watchID = null;

// The radius for our circle object
var radius  = 50;
```

8. Now declare the event listener for the `deviceready` event, as well as the `onDeviceReady` function, which will run once the native PhoneGap code has been loaded:

```
// Set the event listener to run when the device is ready
document.addEventListener("deviceready", onDeviceReady,
false);

// The device is ready so let's
// start watching the acceleration
function onDeviceReady() {

    startWatch();

}
```

9. The `onDeviceReady` function will execute the `startWatch` method, which sets the required `frequency` for accelerometer updates and makes the request to the device to obtain the information:

```
// Watch the acceleration at regular
// intervals as set by the frequency
function startWatch() {

    // Set the frequency of updates from the acceleration
   var options = { frequency: 100 };

    // Assign watchAcceleration to the watchID variable
    // and pass through the options array
    watchID =
        navigator.accelerometer.watchAcceleration(
            onSuccess, onError, options);
}
```

10. With the request made to the device, it is time to create the success and error handling methods. The `onSuccess` function is first, and this will deal with the movement of our object around the screen.
11. To begin, you need to declare some variables that manage the positioning of the element on the device:

```
function onSuccess(acceleration) {

    // Initial X Y positions
    var x = 0;
    var y = 0;

    // Velocity / Speed
    var vx = 0;
    var vy = 0;

    // Acceleration
    var accelX = 0;
    var accelY = 0;

    // Multiplier to create proper pixel measurements
    var vMultiplier =   100;

    // Create a reference to our div elements
    var dot = document.getElementById('dot');
    var accelElement =
    document.getElementById('accelerometerData');
```

```
    // The rest of the code will go here

}
```

12. The returned `acceleration` object contains the information you need regarding the position on the `x` and `y` axis of the device. You can now set the acceleration values for these two axes in the variables and work out the velocity for movement.

13. To correctly interpret the acceleration results into pixels, use the `vMultiplier` variable to convert `x` and `y` into pixels:

```
accelX = acceleration.x;
accelY = acceleration.y;

vy = vy + -(accelY);
vx = vx + accelX;

y = parseInt(y + vy * vMultiplier);
x = parseInt(x + vx * vMultiplier);
```

14. Ensure that the display object doesn't move out of sight and remains within the bounds of the screen:

```
if (x<0) { x = 0; vx = 0; }
if (y<0) { y = 0; vy = 0; }

if (x>document.documentElement.clientWidth-radius) {
  x = document.documentElement.clientWidth-radius; vx = 0;
}

if (y>document.documentElement.clientHeight-radius) {
  y = document.documentElement.clientHeight-radius; vy = 0;
}
```

15. Now that you have the correct `x` and `y` coordinates, you can apply them to the style of the `dot` element position. Also create a string message containing the properties returned from the `acceleration` object as well as the display coordinates that have been created:

```
// Apply the position to the dot element
dot.style.top  = y + "px";
dot.style.left = x + "px";

// Output the acceleration results to the screen
accelElement.innerHTML =
  'Acceleration X: '  + acceleration.x + '<br />' +
  'Acceleration Y: '  + acceleration.y + '<br />' +
  'Acceleration Z: '   + acceleration.z + '<br />' +
```

```
    'Timestamp: '        + acceleration.timestamp +
    '<br />' +
      'Move Top: '             + y + 'px<br />' +
      'Move Left: '            + x + 'px';
```

16. The call to the accelerometer also requires the error handler, so let's write that now. Create a simple string message and insert it into the `div` element to inform the user that a problem has been encountered:

```
// Run if we face an error
// obtaining the accelerometer data
function onError() {

    // Handle any errors we may face
    var accelElement =
            document.getElementById('accelerometerData');

    accelElement.innerHTML =
        'Sorry, I was unable to access the acceleration
        data.';
}
```

17. Finally, add in some CSS to create the `dot` marker used to display the position on the device. Place the following CSS within the `style` tag in the `head` element:

```
<head>
  <style>
  div#dot {
    border-radius: 14px;
    width: 25px;
    height: 25px;
    background: #ff0000;
    position: absolute;
    top: 0px;
    left: 0px;
  }
  </style>
```

18. Upon running the application, you will be able to move the element around the screen by tilting the device. This would look something like this:

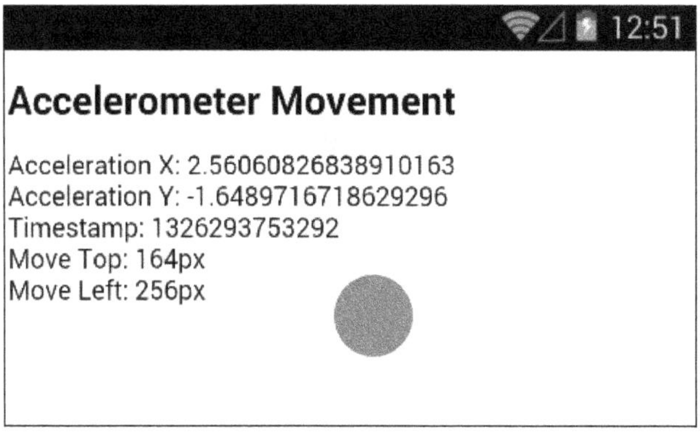

How it works...

By implementing a constant request to watch the acceleration and retrieve movement results from the device, we can pick up changes from the accelerometer sensor. Through some simple JavaScript, we can respond to these changes and update the position of an element around the screen based upon the returned sensor information.

In this recipe, we are easily changing the position of the `dot` element by calculating the correct `x` and `y` axis to place it on the screen. We are also taking extra care to ensure that the element stays within the bounds of the screen by using some conditional statements to check the current position, the radius of the element, and the dimensions of the screen itself.

Obtaining device geolocation sensor information

Geolocation and the use of **Global Positioning System** (**GPS**) allow developers to create dynamic real-time mapping, positioning, and tracking applications. Using the available geolocation methods, we can retrieve a detailed set of information and properties to create location-aware applications. We can obtain the user's location via the mobile data network, Wi-Fi, or directly from the satellite.

How to do it...

We will use the geolocation functionality from the geolocation plugin's API to monitor the feedback from the device and obtain the relevant location information, as follows:

1. First, create a new PhoneGap project named `geolocation`. Open Terminal or Command Prompt and run following command:

   ```
   phonegap create geolocation com.myapp.geolocation geolocation
   ```

2. Add the device platform. You can choose to use Android, iOS, or both:

   ```
   cordova platform add ios
   cordova platform add android
   ```

3. Add the geolocation plugin by running the following command:

   ```
   phonegap plugin add org.apache.cordova.geolocation
   ```

4. Open `www/index.html` and clean up the unnecessary elements; so you have the following:

   ```
   <!DOCTYPE html>
   <html>
       <head>
           <meta charset="utf-8" />
           <meta name="format-detection"
           content="telephone=no" />
           <meta name="msapplication-tap-highlight"
           content="no" />
           <!-- WARNING: for iOS 7, remove the
       width=device-width and height=device-height attributes.
       See https://issues.apache.org/jira/browse/CB-4323 -->
           <meta name="viewport" content="user-scalable=no,
           initial-scale=1, maximum-scale=1, minimum-scale=1,
           width=device-width, height=device-height,
           target-densitydpi=device-dpi" />
           <title>Geolocation</title>
       </head>
       <body>

           <h1>Geolocation Data</h1>

           <div id="geolocationData">Obtaining data...</div>

           <script type="text/javascript"
           src="cordova.js"></script>
           <script type="text/javascript">
   ```

```
        </script>
    </body>
</html>
```

5. Write a new JavaScript tag block beneath the Cordova JavaScript reference. Within this, define an event listener to ensure the device is ready:

```
<script type="text/javascript">
    // Set the event listener to run when the device is
    ready
    document.addEventListener("deviceready", onDeviceReady,
    false);
</script>
```

6. Now add the `onDeviceReady` function. This will execute the `geolocation.getCurrentPosition` method from the PhoneGap API once the native code has fully loaded:

```
// The device is ready so let's
// obtain the current geolocation data
function onDeviceReady() {
    navigator.geolocation.getCurrentPosition(
    onSuccess, onError);
}
```

7. Include the `onSuccess` function to handle the returned the `position` object from the geolocation request.
8. Then, create a reference to the `geolocationData div` element and assign it to the `geoElement` variable, which will hold the generated position results.
9. Next, assign the returned values as a formatted string, which will be set as the HTML content within the `geolocationData div` element. The available properties are accessed through the `position` object:

```
// Run after successful transaction
// Let's display the position data
function onSuccess(position) {

    var geoElement =
    document.getElementById('geolocationData');

    geoElement .innerHTML =
     'Latitude: '  + position.coords.latitude + '<br />' +
     'Longitude: ' + position.coords.longitude + '<br />' +
     'Altitude: '  + position.coords.altitude + '<br />' +
     'Accuracy: '  + position.coords.accuracy + '<br />' +
     'Altitude Accuracy: ' +
     position.coords.altitudeAccuracy + '<br />' +
```

```
        'Heading: ' + position.coords.heading  + '<br />' +
        'Speed: '   + position.coords.speed + '<br />' +
        'Timestamp: ' + position.timestamp + '<br />';
}
```

10. Finally, include the `onError` function to handle any possible errors that may arise.
11. Depending on the existence of an error, use the value of the returned error code to determine which message to display to the user. This will be set as the HTML content of the `geolocationData div` element:

```
// Run if we face an error
// obtaining the position data
function onError(error) {

    var errString   =   '';

    // Check to see if we have received an error code
    if(error.code) {

        // If we have, handle it by case
        switch(error.code)
        {
            case 1: // PERMISSION_DENIED
            errString   =
            'Unable to obtain the location information ' +
            'because the device does not have permission '+
                    'to the use that service.';
            break;
            case 2: // POSITION_UNAVAILABLE
                errString   =
                 'Unable to obtain the location
                 information ' +
                 'because the device location could not ' +
                 'be determined.';
            break;
            case 3: // TIMEOUT
                errString   =
              'Unable to obtain the location within the ' +
                    'specified time allocation.';
            break;
            default: // UNKOWN_ERROR
                errString   =
                  'Unable to obtain the location of the ' +
                    'device due to an unknown error.';
```

```
                break;
            }

        }

        // Handle any errors we may face
        var element =
        document.getElementById('geolocationData');
        element.innerHTML = errString;

    }
```

12. When running the application on a device, the output will look something like this:

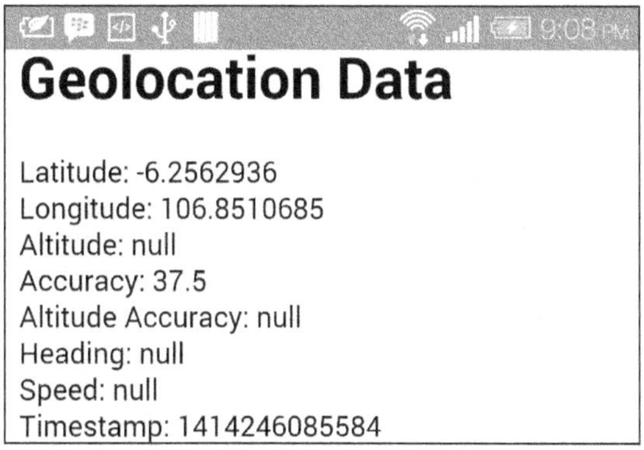

13. If you face any errors, the resulting output will look something like this:

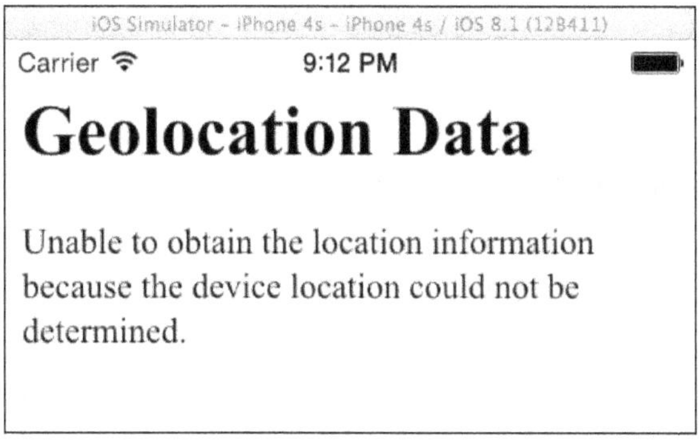

How it works...

As soon as the device is ready and the native PhoneGap code has been initiated on the device, the application will execute the `getCurrentPosition` method from the `geolocation` API. We have defined an `onSuccess` method to manage the output and handling of a successful response, and we have also specified an `onError` method to catch any errors and act accordingly.

The `onSuccess` method returns the obtained geolocation information in the form of the `position` object, which contains the following properties:

- `position.coords.speed`: This is a `Coordinates` object that holds the geographic information returned from the request. This object contains the following properties:
  - `latitude`: This is a `Number` ranging from -90.00 to +90.00 that specifies the latitude estimate in decimal degrees.
  - `longitude`: This is a `Number` ranging from -180.00 to +180.00 that specifies the longitude estimate in decimal degrees.
  - `altitude`: This is a `Number` that specifies the altitude estimate in meters above the **World Geodetic System** (**WGS**) 84 ellipsoid. Optional.
  - `accuracy`: This `Number` specifies the accuracy of the latitude and longitude in meters. Optional.
  - `altitudeAccuracy`: This is a `Number` that specifies the accuracy of the altitude estimate in meters. Optional.
  - `heading`: This is a `Number` that specifies the current direction of movement in degrees, counting clockwise in relation to true north. Optional.
  - `speed`: This `Number` specifies the current ground speed of the device in meters per second. Optional.
- `position.timestamp`: A `DOMTimeStamp` that signifies the time that the geolocation information was received and the `position` object was created.

The properties available within the `position` object are quite comprehensive and detailed. For those marked as *optional*, the value will be set and returned as `null` if the device cannot provide a value.

The `onError` method returns a `PositionError` object if an error is detected during the request. This object contains the following two properties:

- `code`: This is a `Number` that contains a numeric code for the error.
- `message`: This is a `String` that contains a human readable description of the error.

The errors could relate to insufficient permissions needed to access the geolocation sensors on the device, the inability to locate the device due to issues with obtaining the necessary GPS information, a timeout on the request, or the occurrence of an unknown error.

There's more...

The exposed geolocation API accessible through the geolocation plugin is based on the W3C geolocation API specification. Many modern browsers and devices already have this functionality enabled. If any device your application runs on already implements this specification, it will use the built-in support for the API and not the geolocation plugin's implementation.

See also

- You can find out more about geolocation and the `getCurrentPosition` method via the official Cordova documentation available at `http://plugins.cordova.io/#/package/org.apache.cordova.geolocation`.

Adjusting the geolocation sensor update interval

Through the use of the `getCurrentPosition` method, we can retrieve a single reference to the device location using GPS coordinates. In this recipe, we'll create the functionality to obtain the current location based on a numeric interval to receive constantly updated information.

How to do it...

We are able to pass through an optional parameter containing various arguments to set up interval and improve accuracy:

1. First, create a new PhoneGap project named `geoupdate` by running the following command:

   ```
   phonegap create geoupdate com.myapp.geoupdate geoupdate
   ```

2. Add the device platform. You can choose to use Android, iOS, or both:

   ```
   cordova platform add ios
   cordova platform add android
   ```

3. Add the `geolocation` plugin by running the following command:

   ```
   phonegap plugin add org.apache.cordova.geolocation
   ```

4. Open `www/index.html` and clean up unnecessary elements; so you have the following:

```
<!DOCTYPE html>
<html>
    <head>
        <meta charset="utf-8" />
        <meta name="format-detection"
        content="telephone=no" />
        <meta name="msapplication-tap-highlight"
        content="no" />
        <!-- WARNING: for iOS 7, remove the
width=device-width and height=device-height attributes.
See https://issues.apache.org/jira/browse/CB-4323 -->
        <meta name="viewport" content="user-scalable=no,
initial-scale=1, maximum-scale=1, minimum-scale=1,
width=device-width, height=device-height,
target-densitydpi=device-dpi" />
        <title>Geoupdate</title>
    </head>
    <body>
        <h1>Geolocation Data</h1>

        <div id="geolocationData">Obtaining data...</div>

        <script type="text/javascript"
        src="cordova.js"></script>
        <script type="text/javascript">

        </script>
    </body>
</html>
```

5. Below the Cordova JavaScript reference, write a new JavaScript tag block. Within this, declare a new variable called `watchID`.

6. Next, write the event listener to continue once the device is ready:

```
<script type="text/javascript">
    // The watch id variable is set as a
    // reference to the current 'watchPosition'
    var watchID = null;
    // Set the event listener to run
    // when the device is ready
    document.addEventListener("deviceready", onDeviceReady,
    false);
</script>
```

7. Now add the `onDeviceReady` function, which will execute a method called `startWatch`:

```
// The device is ready so let's
// start watching the position
function onDeviceReady() {
    startWatch();
}
```

8. You can now create the `startWatch` function. First, create the `options` variable to hold the optional parameters that can be passed through to the method. Set `frequency` to 5,000 milliseconds (five seconds) and set `enableHighAccuracy` to `true`.
9. Next, assign the `watchPosition` to the previously defined variable `watchID`. This variable will be used to check if the location is currently being watched.
10. To pass through the extra parameters that have been set, send the `options` variable into the `watchPosition` method:

```
function startWatch() {

    // Create the options to send through
    var options = {
            enableHighAccuracy: true
    };

    // Watch the position and update
    // when a change has been detected
    watchID =
navigator.geolocation.watchPosition(onSuccess, onError,
options);

}
```

11. With the initial call methods created, you can now write the `onSuccess` function, which is executed after a successful response. The `position` object from the response is sent through as an argument to the function.
12. Declare some variables to store detailed information obtained from the response in the form of the `timestamp`, `latitude`, `longitude`, and `accuracy` variables. Then, create the element variable to reference the `geolocationData div` element, within which the information will be displayed.
13. The returned information is then assigned to the relevant variables by accessing the properties from the `position` object.

14. Finally, apply the populated variables to a concatenated string and set it as the HTML within the `div` element:

```
// Run after successful transaction
// Let's display the position data
function onSuccess(position) {

    var timestamp, latitude, longitude, accuracy;

    var element =
document.getElementById('geolocationData');

    timestamp   =   new Date(position.timestamp);
    latitude    =   position.coords.latitude;
    longitude   =   position.coords.longitude;
    accuracy        =   position.coords.accuracy;

    element.innerHTML +=
            '<hr />' +
            'Timestamp: '    + timestamp + '<br />' +
            'Latitude: '     + latitude  + '<br />' +
            'Longitude: '    + longitude + '<br />' +
            'Accuracy: '     + accuracy  + '<br />';
}
```

15. With the `onSuccess` method created, now write the `onError` function to handle any errors that you may face following the response:

```
// Run if we face an error
// obtaining the position data
function onError(error) {

    var errString   =   '';

    // Check to see if we have received an error code
    if(error.code) {
        // If we have, handle it by case
        switch(error.code)
        {
            case 1: // PERMISSION_DENIED
                errString   =
            'Unable to obtain the location information ' +
            'because the device does not have permission '+
            'to the use that service.';
            break;
            case 2: // POSITION_UNAVAILABLE
                errString   =
             'Unable to obtain the location information ' +
             'because the device location could not be ' +
             'determined.';
            break;
```

```
            case 3: // TIMEOUT
                errString   =
              'Unable to obtain the location within the ' +
                    'specified time allocation.';
            break;
            default: // UNKOWN_ERROR
                errString   =
                'Unable to obtain the location of the ' +
                'device to an unknown error.';
            break;
        }

    }

    // Handle any errors we may face
    var element =
    document.getElementById('geolocationData');
    element.innerHTML = errString;

}
```

16. Upon running the application, the output will be similar to the following:

How it works...

The `watchPosition` method from the PhoneGap API runs as an asynchronous function, constantly checking for changes to the device's current position. Once a change in position has been detected, it will return the current geographic location information in the form of the `position` object.

With every successful request made on the continuous cycle, the `onSuccess` method is executed, and it formats the data for output onto the screen.

There's more...

There are three optional parameters that can be sent into either the `getCurrentPosition` or `watchPosition` methods, which are:

- `enableHighAccuracy`: This is a `Boolean` that specifies whether or not you would like to obtain the best possible location results from the request. By default (`false`), the position will be retrieved using the mobile or cell network. If set to `true`, more accurate methods will be used to locate the device, using satellite positioning, for example.
- `timeout`: This is a `Number` that defines the maximum length of time in milliseconds to obtain the successful response.
- `maximumAge`: This is a `Number` that defines if a coached position younger than the specified time in milliseconds can be used.

> Android devices will not return a successful geolocation result unless `enableHighAccuracy` is set to `true`.

Clearing the interval

To clear interval timer we can use the `clearWatch` method from the geolocation plugin API. The method to clear the interval and stop watching location data is identical to the method used when clearing accelerometer data obtained from continual updates.

Retrieving map data through geolocation coordinates

In this recipe, we will examine how to render a map on the screen and generate a marker based on latitude and longitude coordinates reported by the device geolocation sensors using the Google Maps API for JavaScript.

Getting ready

Before we can continue with coding the application in this recipe, we must first prepare the project and obtain access to the Google Maps services:

1. First, sign up for a **Google Maps API key**. Visit `https://code.google.com/apis/console/` and log in with your Google account.

2. Select **APIs & auth** followed by **APIs** from the left-hand side menu and activate the **Google Maps JavaScript API v3** service.
3. Once the service has been activated, you must create a new credential. Go to **APIs & auth** | **Credentials**, and under **Public API access** select **Create new Key**. Select **Server key** and hit **Create**. A new key will be created for any IPS.

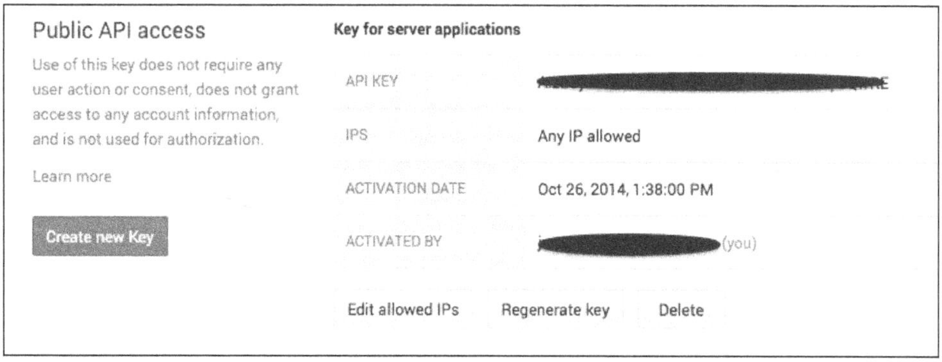

We can now proceed with the recipe.

> `https://developers.google.com/maps/documentation/javascript/.`

How to do it...

We'll use the device's GPS ability to obtain the geolocation coordinates, build and initialize the map canvas, and display the marker for our current position as follows:

1. First, create a new PhoneGap project named `maps` by running the following command:

   ```
   phonegap create maps com.myapp.maps maps
   ```

2. Add the device platform. You can choose to use Android, iOS, or both:

   ```
   cordova platform add ios
   cordova platform add android
   ```

3. Add the geolocation plugin by running the following command:

   ```
   phonegap plugin add org.apache.cordova.geolocation
   ```

4. Open `www/index.html` and clean up unnecessary elements; so you have the following:

   ```
   <!DOCTYPE html>
   <html>
       <head>
           <meta charset="utf-8" />
           <meta name="format-detection"
           content="telephone=no" />
           <meta name="msapplication-tap-highlight"
           content="no" />
           <meta name="viewport" content="user-scalable=no,
           initial-scale=1, maximum-scale=1, minimum-scale=1,
           width=device-width, height=device-height,
           target-densitydpi=device-dpi" />
           <title>You are here</title>
       </head>
       <body>

           <script type="text/javascript"
           src="cordova.js"></script>
           <script type="text/javascript">
               // Custom code goes here
           </script>
       </body>
   </html>
   ```

5. Include the required JavaScript for the Google Maps API under the Cordova reference. Append your API key to the query string in the script `src` attribute:

```
<script type="text/javascript" src="http://maps.googleapis.com/
maps/api/js?key=your_api_key&sensor=true">
</script>
```

> When we included the Google Maps API JavaScript into our document, we set the `sensor` query parameter to `true`. If we were only allowing the user to manually input coordinates without automatic detection, this could have been set to `false`. However, we are using the data obtained from the device's sensor to automatically retrieve our location.

6. Start creating the custom code within the JavaScript tag block. First, create the event listener to ensure the device is ready and also create the `onDeviceReady` method, which will run using the listener:

```
// Set the event listener to run when the device is ready
document.addEventListener(
"deviceready", onDeviceReady, false);

// The device is ready, so let's
// obtain the current geolocation data
function onDeviceReady() {
    navigator.geolocation.getCurrentPosition(
        onSuccess,
        onError
    );
}
```

7. Next, write the `onSuccess` method, which will give you access to the returned location data via the `position` object.
8. Take the `latitude` and `longitude` information obtained from the device geolocation sensor response and create a `latLng` object to be sent into the `Map` when the component is initialized.
9. Then set the options for the `Map`, setting its center to the coordinates set into the `latLng` variable. Not all of the Google Maps controls translate well to the small screen, especially in terms of usability. You can define which controls you would like to use. In this case, accept `zoomControl` but not `panControl`.
10. To define the `Map` itself, reference a `div` element and pass it through the `mapOptions` previously declared.

11. To close this method, create a `Marker` to display at the exact location as set in the `latLng` variable:

```
// Run after successful transaction
// Let's display the position data
function onSuccess(position) {

    var latLng  =
            new google.maps.LatLng(
                    position.coords.latitude,
                    position.coords.longitude);

    var mapOptions = {
                center: latLng,
                panControl: false,
                zoomControl: true,
                zoom: 16,
                mapTypeId: google.maps.MapTypeId.ROADMAP
            };

    var map = new google.maps.Map(
                document.getElementById('map_holder'),
                            mapOptions
                        );

    var marker = new google.maps.Marker({
                position: latLng,
                map: map
            });

}
```

12. To ensure that you correctly handle any errors that may occur, include the `onError` function, which will display the specific string message according to the error within a `div` element:

```
// Run if we face an error
// obtaining the position data
function onError(error) {

    var errString   =   '';

    // Check to see if we have received an error code
    if(error.code) {
        // If we have, handle it by case
```

```
        switch(error.code)
        {
            case 1: // PERMISSION_DENIED
                errString   =
                    'Unable to obtain the location
information ' +
                    'because the device does not have
permission '+
                    'to the use that service.';
            break;
            case 2: // POSITION_UNAVAILABLE
                errString   =
                    'Unable to obtain the location
information ' +
                    'because the device location could not
be ' +
                    'determined.';
            break;
            case 3: // TIMEOUT
                errString   =
                    'Unable to obtain the location within
the ' +
                    'specified time allocation.';
            break;
            default: // UNKOWN_ERROR
                errString   =
                    'Unable to obtain the location of the '
+
                    'device due to an unknown error.';
            break;
        }

    }

    // Handle any errors we may face
    var element = document.getElementById('map_holder');
    element.innerHTML = errString;
}
```

13. With the `body` tag, include the `div` element into which the map will be displayed:

```
<body>
    <div id="map_holder"></div>
</body>
```

14. Finally, add a `style` block within the `head` tag to supply some essential formatting to the page and the `map` element:

```
<style type="text/css">
    html { height: 100% }
    body { height: 100%; margin: 0; padding: 0 }
    #map_holder { height: 100%; width: 100%; }
</style>
```

15. Upon running the application on the device, you will see a result similar to the following:

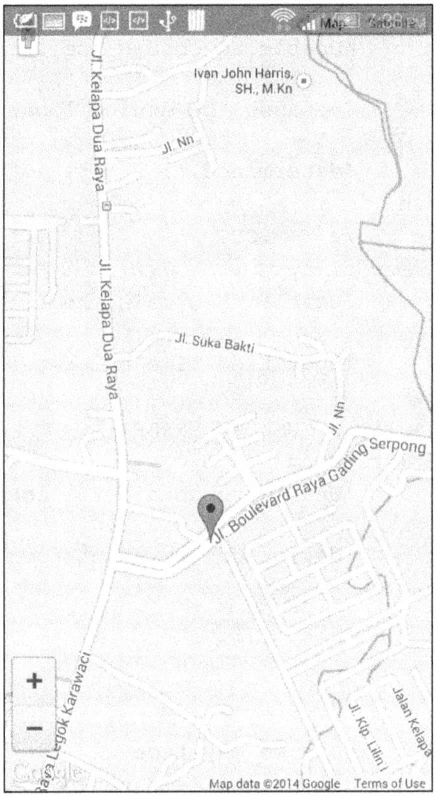

How it works...

Thanks to the use of exposed mapping services such as Google Maps, we are able to perform geolocation updates from the device and use the obtained data to create rich, interactive visual mapping applications.

In this example, we centered the `Map` using the device coordinates and also created a `Marker` overlay to place upon the mark for easy visual reference.

The available APIs for mapping services such as this are incredibly detailed and contain many functions and methods to assist you in creating your location-based tools and applications. Some services also set limits on the number of requests made to the API, so make sure you are aware of any restrictions in place.

Static maps

In this recipe, we used the dynamic Google Maps API. We did this so that we could use the zoom controls and provide our user with a certain level of interaction by being able to drag the map. As an alternative, you could use the Google Static Maps service, which simplifies the code needed to generate a map and will return a static image showing the location.

You can choose to use an API key with this service, but it is not required. You will still have to enable the API in the same way the API access was enabled at the start of this recipe.

Consider the following code. It is an amendment to the `onSuccess` method, which runs after the geolocation data has been obtained:

```
// Run after successful transaction
// Let's display the position data
function onSuccess(position) {
    var mapOutput = '<img
src="http://maps.googleapis.com/maps/api/staticmap?center='+position.
coords.latitude+','+position.coords.longitude+'&zoom=12&size=300x300&s
cale=2&sensor=true">';
    var element = document.getElementById('map_holder');
    element.innerHTML = mapOutput;
}
```

Here, instead of creating the coordinates, the map, and the markers as in the earlier code listing, we simply request an image source using the Static Maps API, and send in the coordinates, image size, and other data as parameters.

By using the Static Maps API, you lose the interactivity offered through the dynamic map, but you gain an incredibly simple, easy-to-use service that requires very little code to achieve results.

There's more...

We used the Google Maps API for JavaScript in this recipe. There are variations in the API level offered by Google, and other mapping systems are also available through other providers, such as MapQuest, MultiMap, and Yahoo! Maps. Explore the alternatives and experiment to see if a particular solution suits your application better than others.

See also

- You can find out more about the Google Maps API from the official documentation at `https://developers.google.com/maps/documentation/javascript/`.
- You can also find out more about the Google Static Maps API on the official documentation, available at `https://developers.google.com/maps/documentation/staticmaps/`.

Creating a visual compass to show the device direction

The geolocation and accelerometer plugins' API provides developers with the ability to receive coordinate and heading information from the device. We can use this information to build a custom compass tool that responds to the device movement.

How to do it...

1. First, create a new PhoneGap project named `compass` by running the following command:

   ```
   phonegap create compass com.myapp.compass compass
   ```

2. Add the device platform. You can choose to use Android, iOS, or both:

   ```
   cordova platform add ios
   cordova platform add android
   ```

3. Add the `device-orientation`, `device-motion`, and `geolocation` plugins by running the following command:

   ```
   phonegap plugin add org.apache.cordova.device-motion
   phonegap plugin add org.apache.cordova.geolocation
   phonegap plugin add org.apache.cordova.device-orientation
   ```

4. Open `www/index.html` and clean up unnecessary elements; so you have the following:

   ```
   <!DOCTYPE html>
   <html>
       <head>
           <meta charset="utf-8" />
           <meta name="format-detection"
           content="telephone=no" />
           <meta name="msapplication-tap-highlight"
           content="no" />
   ```

```
        <meta name="viewport" content="user-scalable=no,
        initial-scale=1, maximum-scale=1, minimum-scale=1,
width=device-width, height=device-height, target-
densitydpi=device-dpi" />
        <title>Compass</title>
    </head>
    <body>

        <script type="text/javascript"
        src="cordova.js"></script>
        <script type="text/javascript">
            // further code will be here
        </script>
    </body>
</html>
```

5. In this example, certain elements within the DOM will be referenced by class name. For this, the **XUI** JavaScript library (`http://xuijs.com/`) will be used. Add the `script` reference under the `cordova` reference:

```
<script type="text/javascript" src="cordova.js"></script>
<script type="text/javascript" src="xui.js"></script>
```

6. Add a new `div` element within the `body` tag and give this the `class` attribute of `container`. This will hold the compass elements for display.
7. The compass itself will be made up of two images. Both images will have an individual `class` name assigned to them, which will allow you to easily reference each of them within the JavaScript. Add these two within the `container` element.
8. Next, write a new `div` element below the images with the `id` attribute set to `heading`. This will hold the text output from the compass response:

```
<body>
    <div class="container">

            <img src="images/rose.png" class="rose"
width="120" height="121" alt="rose" />

            <img src="images/compass.png" class="compass"
width="200" height="200" alt="compass" />

        <div id="heading"></div>

    </div>
```

9. With the initial layout complete, start writing the custom JavaScript code. First, define the `deviceready` event listener—as `XUI` is being used, this differs a little from other recipes within this chapter:

```
var headingDiv;
x$(document).on("deviceready", function () {
});
```

10. When you have a result to output to the user of the application, you want the data to be inserted into the `div` tag with the `heading` `id` attribute. `XUI` makes this a simple task; so update the `headingDiv` global variable to store this reference:

```
x$(document).on("deviceready", function () {
    headingDiv = x$("#heading");
});
```

11. Now include the requests to the PhoneGap compass methods. We'll actually call two within the same function. First, obtain the current heading of the device for instant data, and then make a request through to watch the device heading, making the request every tenth of a second by using the `frequency` parameter; this will provide continual updates to ensure the compass is correct:

```
navigator.compass.getCurrentHeading(onSuccess, onError);
navigator.compass.watchHeading(onSuccess, onError,
{frequency: 100});
```

12. Both of these requests use the same `onSuccess` and `onError` method to handle output and data management. The `onSuccess` method will provide the returned data in the form of a `heading` object.

13. You can use this returned data to set the HTML content of the heading element with the generated message string, using the `headingDiv` variable defined earlier.

14. The visual compass also needs to respond to the heading information. Using `XUI`'s `CSS` method, you can alter the `transform` properties of the rose image to rotate using the returned `magneticHeading` property. Here, reference the image by calling its individual class name, `.rose`:

```
// Run after successful transaction
// Let's display the compass data
function onSuccess(heading) {
    headingDiv.html(
        'Heading: ' + heading.magneticHeading + '&#xb0; ' +
        convertToText(heading.magneticHeading) + '<br />' +
        'True Heading: ' + heading.trueHeading + '<br />' +
        'Accuracy: ' + heading.headingAccuracy
    );
```

```
        // Alter the CSS properties to rotate the rose image
        x$(".rose").css({
            "-webkit-transform":
            "rotate(-" + heading.magneticHeading + "deg)",
           "transform":
            "rotate(-" + heading.magneticHeading + "deg)"
        });

    }
```

15. With the onSuccess handler in place, you now need to add the onError method to output a user-friendly message should you encounter any problems obtaining information:

```
    // Run if we face an error
    // obtaining the compass data
    function onError() {
        headingDiv.html(
            'There was an error trying to ' +
            'locate your current bearing.'
        );
    }
```

16. When creating the message string in the onSuccess function, you made a call to a new function called convertToText. This accepts the magneticHeading value from the heading object and converts it into a text representation of the direction for display. Include this function outside the XUI deviceready block:

```
    // Accept the magneticHeading value
    // and convert into a text representation
    function convertToText(mh) {
        var textDirection;
        if (typeof mh !== "number") {
            textDirection = '';
        } else if (mh >= 337.5 || (mh >= 0 &&  mh <= 22.5)) {
            textDirection =  'N';
        } else if (mh >= 22.5 && mh <= 67.5) {
           textDirection =  'NE';
        } else if (mh >= 67.5 && mh <= 112.5) {
           textDirection =  'E';
        } else if (mh >= 112.5 && mh <= 157.5) {
           textDirection =  'SE';
        } else if (mh >= 157.5 && mh <= 202.5) {
           textDirection =  'S';
        } else if (mh >= 202.5 && mh <= 247.5) {
           textDirection =  'SW';
```

```
        } else if (mh >= 247.5 && mh <= 292.5) {
          textDirection =  'W';
        } else if (mh >= 292.5 && mh <= 337.5) {
          textDirection =  'NW';
        } else {
           textDirection =  textDirection;
        }
        return textDirection;
    }
```

17. Now provide some CSS to position the two images on the screen and ensure the rose image is overlaying the compass image. Add a new `<style>` tag in `<head>` before the `<title>` tag:

```
<style type="text/css">
    .container {
        position: relative;
        margin: 0 auto;
        width: 200px;
        overflow: hidden;
    }

    #heading {
        position: relative;
        font-size: 24px;
        font-weight: 200;
        text-shadow: 0 -1px 0 #eee;
        margin: 20px auto 20px auto;
        color: #111;
        text-align: center;
    }
    .compass {
        padding-top: 12px;
    }
    .rose {
        position: absolute;
        top: 53px;
        left: 40px;
        width: 120px;
        height: 121px;
    }
</style>
<title>Compass</title>
```

18. Upon running the application on the device, the output will look something like this:

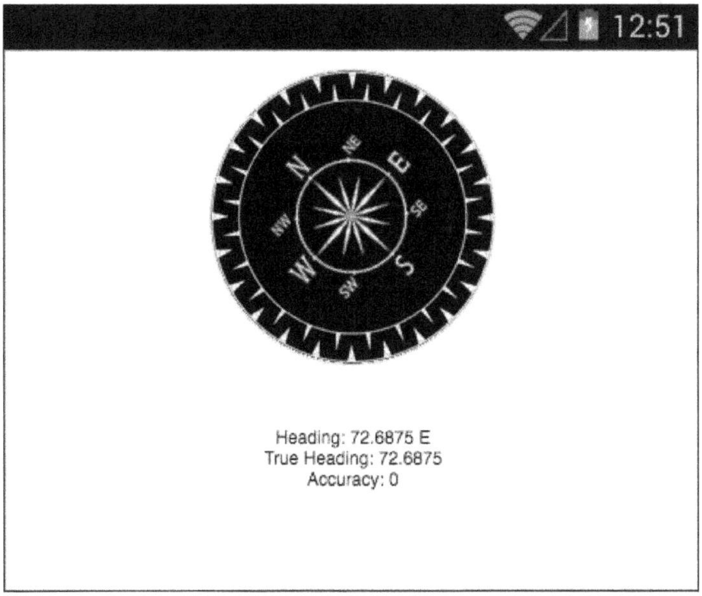

How it works...

The `watchHeading` method from the PhoneGap API compass functionality retrieves periodic updates containing the current heading of the device at the interval specified as the value of the `frequency` variable passed through. If no interval is declared, a default value of 100 milliseconds (one-tenth of a second) is used.

With every successful request made on the continuous cycle, the `onSuccess` method is executed, and it formats the data for output onto the screen as well as making a change to the `transform` property of the graphical element to rotate in accordance with the heading.

The `onSuccess` method returns the obtained heading information in the form of the `compassHeading` object, which contains the following properties:

- `magneticHeading`: This is a `Number` ranging from 0 to 359.99 that specifies a heading in degrees at a single moment in time.
- `trueHeading`: This `Number` ranges from 0 to 359.99 and specifies the heading relative to the geographic North Pole in degrees.
- `headingAccuracy`: This is a `Number` that indicates any deviation in degrees between the reported heading and the true heading values.
- `timestamp`: This refers to the time in milliseconds at which the heading was determined.

See also

- *Chapter 7, Working with XUI*

3

Filesystems, Storage, and Local Databases

In this chapter, we will cover these recipes:

- Saving a file in the device storage
- Opening a local file from the device storage
- Displaying the contents of a directory
- Creating a local SQLite database
- Uploading a file on a remote server via a POST request
- Caching content using the local storage API

Introduction

With the ever-increasing storage capacities on offer with each mobile device, whether the storage is built-in or available as an expansion through a card, developers have the ability to interact with and manipulate files stored on the device as well as utilize the API functionality to cache content.

This chapter will explore how we can save and open individual files on the device's local filesystem, create and manage local SQLite databases, upload a local file on a remote server, and cache content using the local storage API.

Saving a file in the device storage

Thanks to the ability to traverse, read, and write to the device filesystem, an application can write a file to either a specific, predefined location or a location chosen by the user within the application.

How to do it...

We will allow the user to enter a remote URL for a file in a textbox to download and save that file in their mobile device:

1. Firstly, create a new PhoneGap project named `storage` by running the following line:

   ```
   phonegap create storage com.myapp.storage storage
   ```

2. Add the devices platform. You can choose to use Android, iOS, or both:

   ```
   cordova platform add ios
   cordova platform add android
   ```

3. Add the `file` and `file-transfer` plugins by running this line:

   ```
   phonegap plugin add org.apache.cordova.file
   phonegap plugin add org.apache.cordova.file-transfer
   ```

4. We're going to use the `XUI` JavaScript library to easily access the DOM elements, so we'll include the reference to the file within the `head` tag.
5. Let's open `www/index.html` and clean up the unnecessary elements, so we will have this:

   ```
   <!DOCTYPE html>
   <html>
       <head>
           <meta charset="utf-8" />
           <meta name="format-detection"
   content="telephone=no" />
           <meta name="msapplication-tap-highlight"
   content="no" />
           <meta name="viewport" content="user-scalable=no,
   initial-scale=1, maximum-scale=1, minimum-scale=1,
   width=device-width, height=device-height, target-
   densitydpi=device-dpi" />
           <link rel="stylesheet" type="text/css"
   href="css/index.css" />
           <title>Storage</title>
           <script type="text/javascript"
   src="js/xui.js"></script>
   ```

```
    </head>
    <body>

        <script type="text/javascript"
src="cordova.js"></script>
        <script type="text/javascript">
        // further code goes here
        </script>
    </body>
</html>
```

6. Within the `body` tag, create two `input` elements. Set the first element's `type` attribute to `text` and the `id` attribute to `file_url`.
7. Set the second `input` element's `type` attribute to `button`, the `id` attribute to `download_btn`, and `value` to equal `Download`.
8. Finally, include a new `div` element and set the `id` attribute to `message`. This will be the container in which our returned output will be displayed:

```
<body>
    <input type="text" id="file_url" value="" />
    <input type="button" id="download_btn" value="Download"
/>

    <div id="message"></div>
    <script type="text/javascript"
src="cordova.js"></script>
</body>
```

9. Within the empty JavaScript tag block, we need to define a global variable called `downloadDirectory`. It will reference the location on the device to store the retrieved file. We'll also add in our event listener for our application, which will run once the native PhoneGap code has been loaded:

```
var downloadDirectory;
document.addEventListener("deviceready", onDeviceReady,
true);
```

10. We can now write our `onDeviceReady` function. The first thing that we need to do is access the root filesystem on the device. Here, we are requesting access to the persistent filesystem. Once a reference has been established, we run the `onFileSystemSuccess` method to continue.
11. We then bind a `click` function to the `download_btn` element using `XUI`, which will run the `download` function when clicked on:

```
function onDeviceReady() {
    window.requestFileSystem(
```

```
            LocalFileSystem.PERSISTENT,
            0,
            onFileSystemSuccess,
            null
        );

        x$('#download_btn').on( 'click', function(e) {
            download();
        });
    }
```

12. With the connection made to the device storage, we can reference the root system using the `fileSystem` object provided by PhoneGap. Here, we then call the `getDirectory` method, providing the name of the directory to gain access to. If it doesn't exist, it will be created for us. After a successful response, the returned `DirectoryEntry` object is assigned to the `downloadDirectory` variable that we set earlier:

```
function onFileSystemSuccess(fileSystem) {
    fileSystem.root.getDirectory('my_downloads',
       {create:true},
            function(dir) {
                downloadDirectory = dir;
            },fail);
}
```

13. Our `download` function will be run when the user clicks on the **Download** button. Firstly, we need to obtain the URL provided by the user from the text input box. We can pass that value through to a new custom method called `getFileName`, which will split the string and return the filename and extension for use later in the function. We can now set a user-friendly message in the `message` container to inform them of our progress.

14. Next, we instantiate a new `FileTransfer` object from the `File-Transfer` plugin API to assist us in downloading the remote object. The `download` method accepts the remote URL to download, the directory on the device to save the file in, and the `success` and `error` callback functions. After a successful operation, we will inform the user of the local path where the file was saved:

```
function download() {
    var fileURL =
document.getElementById('file_url').value;
    var localFileName = getFilename(fileURL);

    x$('#message').html('Downloading ' + localFileName);
```

```
    var fileTransfer = new FileTransfer();
    fileTransfer.download(
        fileURL,
        downloadDirectory.fullPath + '/' + localFileName,
        function(entry){
            x$('#message').html('Download complete. File
saved to: ' + entry.fullPath);
        },
        function(error){
            alert("Download error source " +
JSON.stringify(error));
        }
    );
}
```

15. We include the custom function to obtain the name and extension of the remote file:

```
function getFilename(url) {
    if (url) {
       var m = url.toString().match(/.*\/(.+?)\./);
        if (m && m.length > 1) {
            return m[1] + '.' + url.split('.').pop();
        }
    }
    return "";
}
```

16. Finally, we supply the `fail` method, which is the generic error handler for all our functions within the application:

```
function fail(error) {
    $('#message').html('We encountered a problem: ' +
error.code);
}
```

17. Upon running the application, we can specify a remote file to download on the local storage, and we will be provided with the file's location on the device. The result will look something like this:

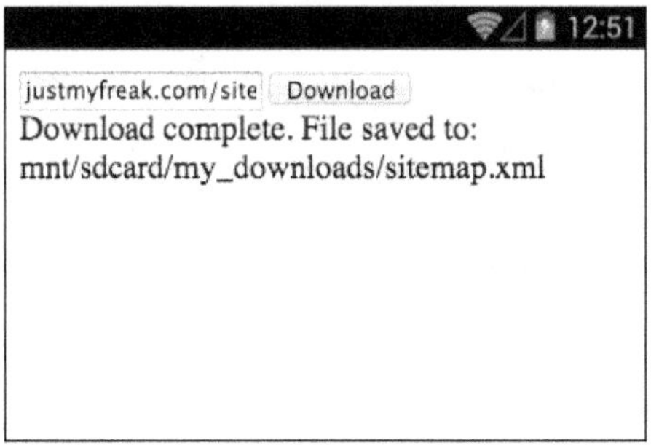

How it works...

In this recipe, we allowed the user to download an external file, publicly accessible on the Internet, and save it at a specified location on the device. Firstly, we needed to create a reference to the `fileSystem` object on the device.

The `fileSystem` object returns the following properties:

- `name`: A `DOMString` object that represents the name of the filesystem
- `root`: A `DirectoryEntry` object that represents the `root` directory of the filesystem

Once obtained, we can obtain the reference to the desired directory location in which our file would be saved using the `DirectoryEntry` object, which returns the following properties:

- `isFile`: A Boolean value that is always `false`, as this is a directory
- `isDirectory`: A Boolean value that is always `true`
- `name`: A `DOMString` representing the name of the directory
- `fullPath`: A `DOMString` that represents the full absolute path of the directory from the root

The `DirectoryEntry` object contains a number of methods that allow you to interact with and manipulate the filesystem. For more information about the available methods, check out the official Cordova file plugin documentation, available at `http://plugins.cordova.io/#/package/org.apache.cordova.file`.

To download the file, we made use of Cordova's `fileTransfer` object from the `file-transfer` plugin, and called the object's `download` method to retrieve the remote file, saving it in the correct directory. You can check out the official documentation of Cordova's `file-transfer` plugin, available at `http://plugins.cordova.io/#/package/org.apache.cordova.file-transfer`.

There's more...

For any Android application that needs to access or write to the device's local storage or filesystem, you have to provide the permission for the application to do so within the `manifest` Android file.

iOS applications also need to have the relevant permissions added to the `Cordova.plist` file to allow access to interact with the device filesystem.

Domain whitelist

One issue that you may encounter while running this example project is an error while trying to download the remote file. Access to remote sites and assets is heavily restricted thanks to the security model in the Cordova project, whereby the default policy is set to block all remote network access.

This can easily be amended by the developer to allow access to specific domains or subdomains, or they can set a wildcard to allow access to every domain, granting full remote network access, by amending the whitelist access specifications.

To find out more about domain whitelists, check out the official documentation at `https://cordova.apache.org/docs/en/3.6.0/guide_appdev_whitelist_index.md.html`.

See also

- *Opening a local file from the device storage*

Opening a local file from the device storage

While developing your mobile application, you may need or want to read particular files from the storage system or from another location on the device.

How to do it...

In this recipe, we will build an application that will create a text file on the phone's storage filesystem, write content to the file, and then open the file to display the content:

1. Firstly, create a new PhoneGap project named `devicestorage` by running this line:

   ```
   phonegap create devicestorage com.myapp.devicestorage devicestorage
   ```

2. Add the devices platform. You can choose to use Android, iOS, or both:

   ```
   cordova platform add ios
   cordova platform add android
   ```

3. Add the `file` plugin by running the following line:

   ```
   phonegap plugin add org.apache.cordova.file
   ```

4. We're going to use the `XUI` JavaScript library to easily access the DOM elements, so we'll include the reference to the file within the `head` tag.
5. Open `www/index.html` and clean up the unnecessary elements. So, we will have this:

   ```
   <!DOCTYPE html>
   <html>
       <head>
           <meta charset="utf-8" />
           <meta name="format-detection"
   content="telephone=no" />
           <meta name="msapplication-tap-highlight"
   content="no" />
           <meta name="viewport" content="user-scalable=no,
   initial-scale=1, maximum-scale=1, minimum-scale=1,
   width=device-width, height=device-height, target-
   densitydpi=device-dpi" />
           <title>Devices Storage</title>
           <script type="text/javascript"
   src="js/xui.js"></script>
       </head>
       <body>
   ```

```
        <script type="text/javascript"
src="cordova.js"></script>
        <script type="text/javascript">
            // further code will be placed here
        </script>
    </body>
</html>
```

6. Within the `body` tag, create two `input` elements. Set the first element's `type` attribute to `text` and the `id` attribute to `my_text`.
7. Set the second `input` element's `type` attribute to `button`, the `id` attribute to `savefile_btn`, and `value` to equal `Save`.
8. Finally, include two new `div` elements. Set the first element's `id` attribute to `message`. This will be the container in which our returned output will be displayed. Set the second element's `id` attribute to `contents`. This will display the contents of the file:

```
<body>
       <input type="text"  id="my_text" />
       <input type="button" id="saveFile_btn" value="Save"
/>

       <div id="message"></div>
       <div id="contents"></div>
</body>
```

9. Within the empty JavaScript tag block, we need to define a global variable called `fileObject`. It will reference the file on the device. We'll also add in our event listener for our application, which will run once the native PhoneGap code has been loaded. The `onDeviceReady` method requests access to the persistent filesystem root on the device. Once obtained, it will execute the `onFileSystemSuccess` method:

```
var fileObject;
document.addEventListener("deviceready", onDeviceReady,
true);

function onDeviceReady() {
    window.requestFileSystem(LocalFileSystem.PERSISTENT, 0,
onFileSystemSuccess, fail);
}
```

> The `LocalFileSystem.PERSISTENT` constant is used here to ensure that we access storage that cannot be removed by the user agent without permission from the application or the user. We could have used the `LocalFileSystem.TEMPORARY` constant to access storage that has no guarantee of persistence, but this is not recommended.

10. With the connection made to the device storage, we can reference the root system using the `fileSystem` object provided by PhoneGap. Here, we then call the `getFile` method, providing the name of the file that we wish to open. If it doesn't exist, it will be created for us:

```
function onFileSystemSuccess(fileSystem) {
    fileSystem.root.getFile("readme.txt",
       {create: true, exclusive: false},
       gotFileEntry, fail);
}
```

11. After a successful response, the returned `FileEntry` object is assigned to the `fileObject` variable we created earlier. At this point, we can also bind a click handler to our **Save** button, which will run the `saveFileContent` function when clicked on:

```
function gotFileEntry(fileEntry) {
    fileObject = fileEntry;
    x$('#saveFile_btn').on('click', function() {
        saveFileContent();
    });
}
```

12. As we're dealing with a local file that doesn't have any content yet, we can use methods within the PhoneGap API to write to the file. The `saveFileContent` method will access `fileObject` and call the `createWriter` method to start this process:

```
function saveFileContent() {
    fileObject.createWriter(gotFileWriter, fail);
}
```

13. Let's now create the `gotFileWriter` method, which is called as a callback, from the `save` function. We'll send the value from the `my_text` input field to the `writer.write()` method to populate the file content. After the writing completes, we'll output a status message in the `message div` element and then instantiate the `FileReader` object to read the file contents.

14. We will then pass `fileObject` to the `reader.readAsText()` method to return the text content of the file. After the read completes, we will output the contents in the `div` element for display:

```
function gotFileWriter(writer) {
    var myText = document.getElementById('my_text').value;
    writer.write(myText);

    writer.onwriteend = function(evt) {
        x$('#message').html('<p>File contents have been
written.<br /><strong>File path:</strong> ' +
fileObject.fullPath + '</p>');
```

```
            var reader = new FileReader();
            reader.readAsText(fileObject);
            reader.onload = function(evt) {
                x$('#contents').html('<strong>File
    contents:</strong> <br />' + evt.target.result);
            };
        };
    }
```

15. Finally, include the `fail` error handler method to catch any problems or errors:

```
    function fail(error) {
        alert(error.code);
    }
```

16. Upon running the application on our device, we should see an output similar to this:

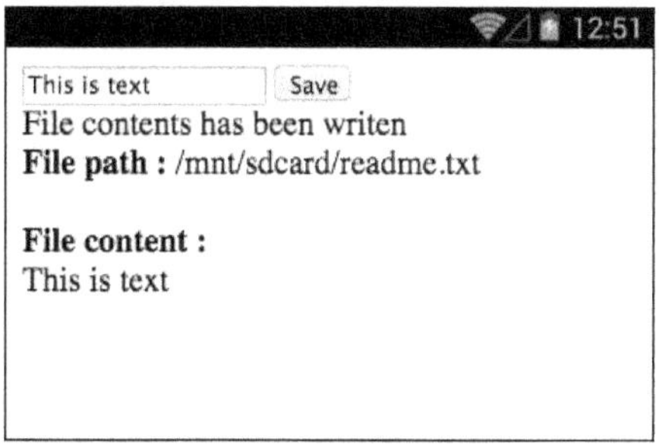

How it works...

To gain access to the filesystem on the device, we first request access to the persistent storage, which provides us with access to the `fileSystem` object.

Navigating to the root of the filesystem, we then call the `getFile` method, which will look up the requested file, or create it if it doesn't already exist in the specified location.

Once a user has typed content in the `input` textbox, we can instantiate a `FileWriter` object on the saved object containing the file reference and write the user-supplied content to the file. We can also make use of the `FileWriter` object's `onwriteend` method, which is called when the request has completed, to output a message to the user and then begin the request to read the contents of the file. This is achieved through the use of the `FileReader` object.

> For a comprehensive look at the `File` functions available within the PhoneGap API, refer to the official documentation at `http://plugins.cordova.io/#/package/org.apache.cordova.file`.

There's more...

In this example, we were able to read the file as text, using the `readAsText` method. We are also able to read a file and return the contents as a base64-encoded data URL using the `readAsDataURL` method. While there are no limitations on what type of file can be read, depending on the choice of reading method as well as the size of the file, consideration must be placed on the impact on speed and performance that may occur while trying to read large files, which may take up quite a lot of processing power.

See also

- *Saving a file in the device storage*

Displaying the contents of a directory

As devices may offer us a lot of storage space that we can potentially use, we can make sure that we have the ability to traverse the filesystem to ascertain the structure of the storage available.

How to do it...

In this recipe, we will build an application that will read the contents of a directory from the device's root filesystem, and display it in a list format:

1. Firstly, create a new PhoneGap project named `showdirectory` by running this line:

   ```
   phonegap create showdirectory com.myapp.showdirectory
   showdirectory
   ```

2. Add the devices platform. You can choose to use Android, iOS, or both:

   ```
   cordova platform add ios
   cordova platform add android
   ```

3. Add the `file` plugin by running the following line:

   ```
   phonegap plugin add org.apache.cordova.file
   ```

4. We're going to use jQuery Mobile. So, we'll include the reference to the JavaScript and CSS files within the `head` tag.
5. Open `www/index.html` and clean up the unnecessary elements. So, we will have this:

   ```
   <!DOCTYPE html>
   <html>
       <head>
   ```

```
        <meta charset="utf-8" />
        <meta name="format-detection"
content="telephone=no" />
        <meta name="msapplication-tap-highlight"
content="no" />
        <meta name="viewport" content="user-scalable=no,
initial-scale=1, maximum-scale=1, minimum-scale=1,
width=device-width, height=device-height, target-
densitydpi=device-dpi" />
        <link rel="stylesheet" href="jquery/jquery.mobile-
1.1.1.min.css" />
        <script src="jquery/jquery-1.8.0.min.js"></script>
        <script src="jquery/jquery.mobile-1.1.1.min.js"></script>
        <title>Show Directory</title>
    </head>
    <body>

        <script type="text/javascript"
src="cordova.js"></script>
        <script type="text/javascript">

        </script>
    </body>
</html>
```

6. The jQuery Mobile framework will handle the formatting and layout of the body content. Include a `div` element with the `data-role` attribute set to `page`. Within this, we add a header to contain our application title. We'll then add a new `div` element with the `data-role` attribute set to `content`, inside which we'll place a `ul` tag block to hold our directory listings. We set `data-role` for the `ul` tag to `listview` and give it an `id` of `directoryList` so that we can reference it later:

```
<body>
  <div data-role="page">

      <div data-role="header">
          <h2>Directory Reader</h2>
      </div>

      <div data-role="content">
          <ul id="directoryList" data-role="listview" data-
inset="true">

          </ul>
      </div>

  </div>
</body>
```

7. Next, we need to add the event listener and the `onDeviceReady` method to run once the native PhoneGap code is ready to be executed. In this method, we will request access to the file root on the persistent storage, which will then run the `onFileSystemSuccess` callback method:

```
document.addEventListener("deviceready", onDeviceReady,
false);
function onDeviceReady(){
    window.requestFileSystem(
       LocalFileSystem.PERSISTENT,
       0, onFileSystemSuccess, fail
    );
}
```

8. To ensure that we have some extra content to list, we'll call the `getDirectory` and `getFile` methods respectively, which will create the directory and file if they do not already exist. We can access the `DirectoryEntry` object at `fileSystem.root`, and call the `createReader()` method from it to instantiate the `DirectoryReader` object. Finally, let's call the `readEntries()` method from this object to read the entries within the provided directory:

```
function onFileSystemSuccess(fileSystem) {
    // Create some test files
    fileSystem.root.getDirectory("myDirectory",
        { create: true, exclusive: false },
        null,fail);
    fileSystem.root.getFile("readthis.txt",
        { create: true, exclusive: false },
        null,fail);

    var directoryReader = fileSystem.root.createReader();
    // Get a list of all the entries in the directory
    directoryReader.readEntries(success,fail);
}
```

9. The `success` callback method is passed as an array of `FileEntry` and `DirectoryEntry` objects. Here, we'll loop over the array and create a list item for each returned entry, writing the name and the URI path. We'll also check the type of the entry and display whether it's a directory or a file.

10. Each list item is appended to the `directoryList ul` element, and we then call a `listview` refresh method on the element to update the content for displaying:

```
function success(entries) {
    var i;
    var objectType;
    for (i=0; i<entries.length; i++) {
```

```
        if(entries[i].isDirectory == true) {
            objectType = 'Directory';
        } else {
            objectType = 'File';
        }
        $('#directoryList').append('<li><h3>' +
entries[i].name + '</h3><p>' + entries[i].toURI() + '</p><p
class="ui-li-aside">Type: <strong>' + objectType +
'</strong></p></li>');
    }
    $('#directoryList').listview("refresh");
}
```

11. Finally, let's include the `fail` error handler method to alert us of any issues, like this:

```
function fail(error) {
    alert("Failed to list directory contents: " +
error.code);
}
```

12. Upon running the application on our device, the output will look something like this:

How it works...

When traversing directories, the PhoneGap API provides the perfect solution in the form of the `DirectoryReader` object, which lists all directories and files within the chosen directory. This contains a single method called `readEntries`, and it's the success callback method from this that allows us to loop over the contents and output them as a visual representation.

See also

- *Chapter 10, User Interface Development, Creating a jQuery Mobile layout*

Creating a local SQLite database

SQLite databases are a fantastic way to store structured information from a web context. SQLite is a self-contained transactional database that does not require any configuration. It is ideal for saving and querying dynamic information within a mobile application.

How to do it...

In this recipe, we will create a mobile application that will allow us to store text entries in a local SQLite database, and then query the database to retrieve all saved items:

1. Firstly, create a new PhoneGap project named `sqlite` by running the following line:

   ```
   phonegap create sqlite com.myapp.sqlite sqlite
   ```

2. Add the devices platform. You can choose to use Android, iOS, or both:

   ```
   cordova platform add ios
   cordova platform add android
   ```

3. Open `www/index.html` and clean up the unnecessary elements. So, this is what you will have:

   ```
   <!DOCTYPE html>
   <html>
       <head>
           <meta charset="utf-8" />
           <meta name="format-detection"
   content="telephone=no" />
           <meta name="msapplication-tap-highlight"
   content="no" />
   ```

```
        <meta name="viewport" content="user-scalable=no,
initial-scale=1, maximum-scale=1, minimum-scale=1,
width=device-width, height=device-height, target-
densitydpi=device-dpi" />
        <title>SQLite</title>
    </head>
    <body>

        <script type="text/javascript"
src="cordova.js"></script>
        <script type="text/javascript">
            // further code goes here
        </script>
    </body>
</html>
```

4. In this example, we will be referencing certain elements within the DOM by class name. To do this, we will use the XUI JavaScript library. Add the script reference within the head tag of the document to include this library.
5. Below the PhoneGap JavaScript include, write a new JavaScript tag block, and within it, define an onDeviceReady event listener to ensure that the device is ready and fully loaded before the application proceeds to execute the code:

```
<script type="text/javascript" src="js/xui.js"></script>
<title>SQLite</title>
```

6. Add an input element within the body tags with the id attribute set to list_action. This will allow the user to add entries to the database list.
7. Below this, add a button element with the id attribute set to saveItem.
8. Let's also add two div elements to hold any generated data. The first, with the id attribute set to message, will hold any database connection error messages if we have issues trying to connect. The second, with the id attribute set to listItems, will act as a container in which our generated list will be placed for display:

```
<body>
  <h1>My ToDo List</h1>

  <input type="text"  id="list_action" />
  <input type="button" id="saveItem" value="Save" />

  <div id="message"></div>
  <div id="listItems"></div>
</body>
```

9. With the layout complete, let's move on to adding our custom JavaScript code. To begin with, we need to define the `deviceready` event listener. As we are using the `XUI` library, for this recipe, we will write this function as follows:

```
x$(document).on("deviceready", function () {

});
```

10. As we want to set the inner HTML values for the `list` and `message div` containers, let's define the references to those particular elements. `XUI` makes this really easy. We'll create a global variable called `db`, which will eventually hold our database connection.
11. Let's also bind a click handler to the `saveItem` button element. When pressed, it will run the `insertItem` method to add a new record to the database.
12. We now need to create a reference to our SQLite database. The PhoneGap API includes a function called `openDatabase` that creates a new database instance or opens the database if it already exists. The returned object will allow us to perform transactions against the database:

```
var listElement      = x$('#listItems');
var messageElement   = x$('#message');
var db;

x$('#saveItem').on('click', function(e) {
    insertItem();
});

// Create a reference to the database
function getDatabase() {
return window.openDatabase("todoListDB",
            "1.0", "ToDoList Database", 200000);
}
```

13. We can now include the call to and create the `onDeviceReady` method. Here, we assign the database instance to the `db` variable, which will allow us to perform a transaction on the database. In this case, we'll execute a simple SQL script to create a table called `MYLIST`, if it doesn't already exist:

```
// Run the onDeviceReady method
onDeviceReady();

// PhoneGap is ready
function onDeviceReady() {
db = getDatabase();
db.transaction(function(tx) {
tx.executeSql('CREATE TABLE IF NOT EXISTS MYLIST
```

```
    (id INTEGER PRIMARY KEY AUTOINCREMENT, list_action)');
    }, databaseError, getItems);
}
```

14. Let's now define the `getItems` method, which is run on a successful callback from the database transaction in the previous method. Once again, we reference the database object and perform another transaction, this time to select all items from the table:

```
function getItems() {
    db.transaction(function(tx) {
    tx.executeSql('SELECT * FROM MYLIST', [],
       querySuccess, databaseError);
    }, databaseError);
}
```

15. Having received the results from the select query, we can loop over them and create list item elements that we can then set within the list `div` container. Here, we can reference the `id` and `list_action` columns from the results, drawn from the SQLite database table that we created earlier. We'll also output a user-friendly message displaying the total number of records stored, as follows:

```
// Process the SQLResultSetList
function querySuccess(tx, results) {
    var len = results.rows.length;
    var output = '';
    for (var i=0; i<len; i++){
       output = output +
          '<li id="' + results.rows.item(i).id + '">' +
          results.rows.item(i).list_action + '</li>';
    }
    messageElement.html('<p>There are ' + len + ' items in
your list:</p>');
    listElement.html('<ul>' + output + '</ul>');
}
```

16. We initially bound a click event handler to our `saveItem` button. Let's now create the `insertItem` method, which the `click` handler will invoke. We want to take the value of the `list_action` text input box and pass it to the database transaction when we run an insert query. A successful insertion will call our `getItems` method to query the database and populate the list with all updated information from our database:

```
// Insert a record into the database
function insertItem() {
    var insertValue =
    document.getElementById('list_action').value;
```

```
    db.transaction(function(tx) {
    tx.executeSql('INSERT INTO MYLIST
       (list_action) VALUES ("' + insertValue + '")');
    }, databaseError, getItems);
    // Clear the value from the input box
    document.getElementById('list_action').value = '';
}
```

17. Finally, let's include our `databaseError` fault handler method to display any issues that we may encounter from the database, and display them in the `message div` element:

```
// Database error handler
function databaseError(error) {
    messageElement.html("SQL Error: " + error.code);
}
```

18. Upon running the application on the device, your output should look something like what is shown here:

How it works...

In order to access the SQLite database and perform any transactions, we first need to establish a connection with the `.db` file on the device using the `openDatabase` method. Once this connection is established, we can use the `SQLTransaction` object to perform `executeSql` methods such as table creation, selection, and insertion queries, written using standard SQL syntax.

> For more details on the full methods available for use with the SQLite implementation, check out the official Cordova storage documentation at `http://docs.phonegap.com/en/3.5.0/cordova_storage_storage.md.html`.

There's more...

The storage API accessible through Cordova is based on the W3C Web SQL Database specification. Some devices already have an implementation of this specification. If any device that your application runs on provides this functionality, it will use its built-in support for the storage specification and will not use Cordova's implementation.

See also

- Chapter 7, Working with XUI

Uploading a file on a remote server

Sometimes, working with only the local system for mobile applications is not enough. There are use cases for the requirement to interact with a remote server, for example, to share a file.

How to do it...

In this recipe, we will build an application that allows the user to take a photograph and upload it on a remote server:

1. Firstly, create a new PhoneGap project named `upload` by running the following line:

   ```
   phonegap create upload com.myapp.upload upload
   ```

2. Add the devices platform. You can choose to use Android, iOS, or both:

   ```
   cordova platform add ios
   cordova platform add android
   ```

3. Add the `file`, `camera`, and `file-transfer` plugins by running these lines of code:

   ```
   phonegap plugin add org.apache.cordova.file
   phonegap plugin add org.apache.cordova.camera
   phonegap plugin add org.apache.cordova.file-transfer
   ```

4. We're going to use the `XUI` JavaScript library to easily access the DOM elements, so we'll include the reference to the file within the `head` tag.
5. Open `www/index.html`. Let's clean up the unnecessary elements, so we will have this:

   ```
   <!DOCTYPE html>
   <html>
       <head>
           <meta charset="utf-8" />
   ```

```
        <meta name="format-detection"
content="telephone=no" />
        <meta name="msapplication-tap-highlight"
content="no" />
        <meta name="viewport" content="user-scalable=no,
initial-scale=1, maximum-scale=1, minimum-scale=1,
width=device-width, height=device-height, target-
densitydpi=device-dpi" />
        <script type="text/javascript"
src="js/xui.js"></script>
        <title>Upload</title>
    </head>
    <body>

        <script type="text/javascript"
src="cordova.js"></script>
        <script type="text/javascript">

        </script>
    </body>
</html>
```

6. We now need to create some elements within the `body` tag. Let's include a `button` element with the `id` attribute set to `selectorBtn`. Create a `div` element with the `id` attribute set to `message`, which we'll use to display status updates from the transfer. Finally, we'll create an `img` tag with the `id` attribute set to `returnImage`:

```
<body>
    <button id="selectorBtn">Take Photo</button>

    <div id="message"></div>

    <img id="returnImage" />
</body>
```

7. We can now include our event handler to ensure that the device is ready before proceeding, as well as create our `onDeviceReady` method. In this method, we will bind the `selectorBtn` element to a `touchstart` event. When pressed, this will call the `camera.getPicture` method from the PhoneGap API, which will open the device's default camera application to let our user take a photo to upload:

```
document.addEventListener("deviceready", onDeviceReady,
true);

function onDeviceReady() {

    x$("#selectorBtn").touchstart(function(e) {
```

```
            navigator.camera.getPicture(
                gotPicture,
                    onError,
                    {
                    sourceType:Camera.PictureSourceType.CAMERA,
               destinationType:Camera.DestinationType.FILE_URI,
                        quality:50
                    }
            );
        });
    }
```

Use `Camera.PictureSourceType.PHOTOLIBRARY` or `Camera.PictureSourceType.SAVEDPHOTOALBUM` if we wanted the user to select an image to upload from their saved photos.

1. We now include the `gotPicture` method as the success callback, having obtained an image. We have the location of the file as a provided parameter. Let's update the `message div` to display a friendly message to our users, and then we'll create a new instance of the `FileUploadOptions` object, which we'll use to specify additional parameters for the upload script. The `options.fileKey` value sets the name for the form field that will contain the uploaded file.
2. We can now create a new instance of the `FileTransfer` object, from which we'll call the `upload` method. Here, we can pass the file's location on the device and the remote address to upload it at, and also send any additional parameters that we may have included in the `options` object:

```
function gotPicture(fileLocation) {
    x$("#message").html("<p>Uploading your image...</p>");

    var options = new FileUploadOptions();
    options.fileKey     =   "file";
    options.fileName    =
fileLocation.substr(fileLocation.lastIndexOf('/')+1);
    options.mimeType    =   "image/jpeg";
    options.chunkedMode =   false;

    var fileTransfer = new FileTransfer();
    fileTransfer.upload(
        fileLocation,
        "http://address_to_remote_server_page/upload.php",
        fileUploaded,
        onError,
        options
    );
}
```

> In this example, I have used ColdFusion as the dynamic server-side language to process the upload. You can, of course, use any server-side language that you have access to or feel comfortable using to manage the upload.

3. Let's now create the callback handler, following a successful upload. The `response` parameter is a `FileUploadResult` object, and from it, we can obtain the total number of bytes sent as well as the output response from the server. In this case, we are returning the image from the server-side script, and we will set it as the `src` attribute for the `returnImage img` element, like this:

```
function fileUploaded(result) {
x$("#message").html('<p>Upload complete!!<br />Bytes sent:
' + result.bytesSent + '</p>');
    x$("#returnImage").attr("src", result.response);
}
```

4. Finally, let's create our `onError` fault handler to alert us of any possible issues:

```
function onError(error) {
    alert("Error: " + JSON.stringify(error));
}
```

5. When you run the application on a device, the output should look something like this:

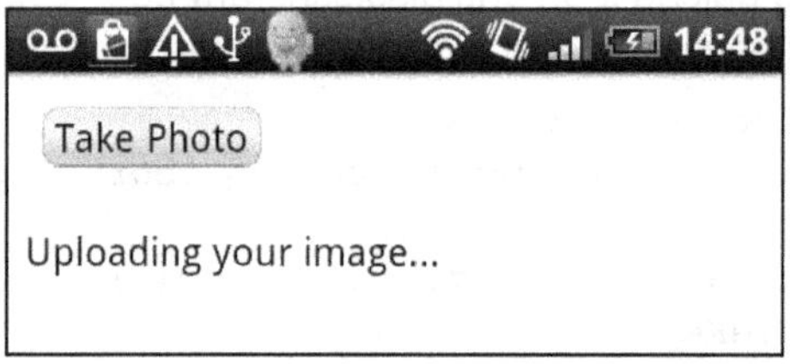

6. Following a successful response from the remote server, the application will display something similar to this:

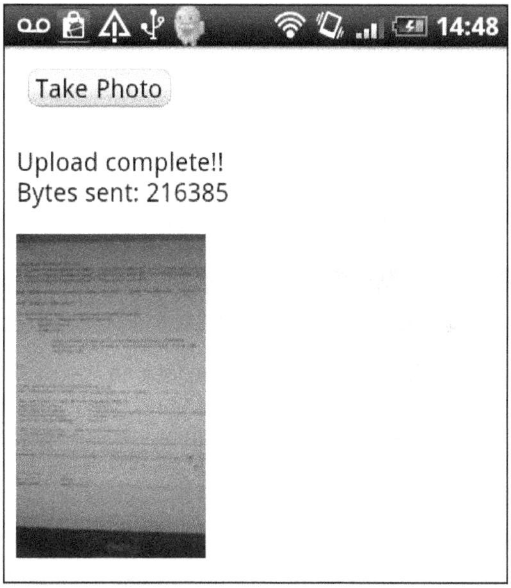

How it works...

Firstly, we need to define how we are going to retrieve our image from the device using the `camera.getPicture` method. Once we have obtained our picture, we can start building our `FileTransfer` object, which will handle the transaction to the remote server for us. We can then create a `FileUploadOptions` object, which specifies any additional parameters to use on the server-side handling page. The properties available for use within the `FileUploadOptions` object are as follows:

- `fileKey`: A `DOMString` that represents the name of the `form` element. The default value is `file`.
- `fileName`: A `DOMString` that represents the name you wish the file to be saved as on the server. The default value is `image.jpg`.
- `mimeType`: A `DOMString` that represents the MIME type of the data that you wish to upload. The default value is `image/jpeg`.
- `params`: An object that allows you to set optional key/value pairs to be included in the HTTP request.
- `chunkedMode`: A boolean value that determines whether or not the data should be uploaded in chunked streaming mode. The default value is `true`.

Finally, our success callback method will contain the `FileUploadResult` object returned from the transaction, which gives us access to the response from the server as well as the number of bytes sent in the upload, which we can then output, store, or use in any way we need to.

For more details on the `FileTransfer` object, check out the official Cordova documentation at `http://docs.phonegap.com/en/2.0.0/cordova_file_file.md.html#FileTransfer`.

See also

- Chapter 6, *Hooking into Native Events, Displaying network connection status*

Caching content using the local storage API

As mobile users access applications and pull remote data on the move, we need to be conscious and aware that our application may be using up the limited data services. We can implement services and techniques to help reduce unnecessary remote calls to data.

How to do it...

In this recipe, we will build an application that allows the user to search Twitter using its open API. We'll store the results for the search in `localStorage` so that they are available when we reopen the application:

1. Firstly, create a new PhoneGap project named `localstorage` by running this line:

   ```
   phonegap create localstorage com.myapp.localstorage localstorage
   ```

2. Add the devices platform. You can choose to use Android, iOS, or both:

   ```
   cordova platform add ios
   cordova platform add android
   ```

3. We will be using the jQuery Mobile framework for our layout, so we include the relevant CSS and JavaScript file references within the `head` tag.
4. Open `www/index.html` and clean up the unnecessary elements. So, we will have this code:

   ```
   <!DOCTYPE html>
   <html>
       <head>
           <meta charset="utf-8" />
           <meta name="format-detection"
   content="telephone=no" />
           <meta name="msapplication-tap-highlight"
   content="no" />
           <meta name="viewport" content="user-scalable=no,
   initial-scale=1, maximum-scale=1, minimum-scale=1,
   width=device-width, height=device-height, target-
   densitydpi=device-dpi" />
   ```

```
        <link rel="stylesheet" href="jquery/jquery.mobile-
1.1.1.min.css" />
        <script src="jquery/jquery-1.8.0.min.js"></script>
        <script src="jquery/jquery.mobile-1.1.1.min.js"></script>
        <title>LocalStorage</title>
    </head>
    <body>

        <script type="text/javascript"
src="cordova.js"></script>
        <script type="text/javascript">

        </script>
    </body>
</html>
```

5. Within the `body` tag, create a new `div` element with the `data-role` attribute set to `page`, which will form the container for the jQuery Mobile layout:

```
<body>
    <div data-role="page">


     </div>
</body>
```

6. Let's place some more layout structure in our application, which is required by the jQuery Mobile framework. Create a new `div` element with the `id` attribute set to `header`. The `data-role` and `data-position` attributes must also be set as shown.
7. Within the header, we'll display an `anchor` tag to exit the application. We specify the `id` attribute and also set a specific icon to display in the header thanks to the `data-icon` attribute.
8. We'll include an `h2` heading to add a title to the application, as well as a second button, which we'll use to clear any cached content. This too has a specific icon set in the `data-icon` attribute as well as the `id` attribute to allow us to reference it via our JavaScript:

```
<div id="header" data-role="header" data-position="inline">

    <a id="exit_btn" data-inline="true"  data-theme="b"
data-icon="home">Exit</a>

    <h2>Local Storage Search</h2>
```

```
    <a id="clear_btn" data-inline="true" data-theme="b"
data-icon="delete">Clear Storage</a>

</div>
```

9. Below this, create a new `div` element with the `data-role` attribute set to `content`. This will house a second `div` block, inside which we'll place two `input` elements. The first holds the user's search criteria, the second is the button used to perform the search.
10. We'll now include a `ul` element with the `id` attribute set to `tweetResults` and the `data-role` attribute set to `listview`, which will hold our returned data:

```
<div data-role="content">

    <div data-role="fieldcontain">

        <input type="search" name="search" id="searchTerm" data-
inline="true" data-icon="search" />
        <input type="button" id="search_btn" value="Search" data-
theme="b" data-inline="true" />

    </div>

    <ul id="tweetResults" data-role="listview"
        data-inset="true">

    </ul>

</div>
```

11. With the layout complete, we can start adding our custom code. Include a new JavaScript tag block before the closing `head` tag. Inside it, let's write an event listener to ensure that the PhoneGap native code has loaded before we proceed:

```
document.addEventListener("deviceready", onDeviceReady,
true);
```

12. Let's start adding the custom code that will be run when the device is ready. Create the `onDeviceReady` function. At the start, we'll set some required variables; the first two will hold some messages to output to our users. The second is a reference to the `localStorage` interface, which we'll use to save our data in key-value pairs.
13. We need to run a check to see whether any content from a previous request has been stored by calling the `getItem` method on the `localStorage` object. If it exists, we'll display a user-friendly message, set the search term from the previous search in the input box, display the `clear` button by calling the `showClearButton` method, and finally loop over the results to display them by calling the `outputResults` method.

14. If we have no previous results saved in `localStorage`, we'll simply display a welcome message to the user and ensure that the clear button is hidden:

```
function onDeviceReady() {
    // Create the friendly messages and define the
variables
    var previousMessage =   'Here are your previous search
results..';
    var welcomeMessage  =   'What would you like to search
for?';
    var localStorage    =   window.localStorage;

    /* Firstly, check to see if localStorage
    has any cached content from a previous request. */
    if(localStorage.getItem('twitSearchResults')) {

        /* We have saved content,
     so display a nice message to the user */
        $('body h2').html(previousMessage);

        /* Set the value of the stored search
     term into the input box */
  $('#searchTerm').val(localStorage.getItem('searchTerm'));

        // Display the clear button
        showClearButton();

        // Send the stored data to be rendered as HTML
outputResults(JSON.parse(localStorage.
getItem('twitSearchResults')));
    } else {
        /* There is nothing cached,
        so display a friendly message */
        $('body h2').html(welcomeMessage);
        hideClearButton();
    }

    // add click handlers here

}
```

15. Within the `onDeviceReady` function, we now need to set up the `click` handlers for each of our buttons. The first is the `clear_btn` element, which when clicked on will clear the values in the `tweetResults div` element, and remove the data that we have cached by calling the `clear` method on the `localStorage` object.

16. The second handler is applied to the `exit_btn` element, which will gracefully close the application:

```
/* Add a click handler to the clear button
which will be displayed is a user returns
to the page with saved content */
$('#clear_btn').click(function() {
    // Clear the entire local storage object
    localStorage.clear();
    // Clear the content list
    $('#tweetResults').html('');
    $('#tweetResults').hide();

    /* There is nothing cached,
     so display a friendly message */
    $('body h2').html(welcomeMessage);

    // Remove the clear button
    hideClearButton();

    // Reset the search term input field
    $('#searchTerm').val('');

});

$('#exit_btn').click(function() {
    navigator.app.exitApp();
});
```

17. The third click handler will be placed on the `search_btn` element, which will take the search term provided by the user and pass it to a new function, called `makeSearchRequest`:

```
/* Add a click handler to the search button
    which will make our AJAX requests for us */
$('#search_btn').click(function() {

/* Obtain the value of the search term and send it
    through to the request function */
makeSearchRequest($('#searchTerm').val());

});
```

18. Before we make the call, we'll save the search term in `localStorage` using the `setItem()` method. This will allow us to reference it at any time until we have cleared the cache.

19. To make the request to the remote API, we'll utilize jQuery's built-in `ajax()` method, here asking for five results per page on the search term provided by the user. To handle the returned data, we'll also specify the `jsonpCallback` method, which in this case is a new function called `storeResults`:

```
function makeSearchRequest(searchTerm) {

    // Display a user-friendly message
    $('body h2').html('Searching for: '+ searchTerm);

    /* Store the value we are searching
    for into the localStorage object */
    localStorage.setItem('searchTerm', searchTerm);

    // Make the request to the Twitter search API
    $.ajax({
        url: "http://search.twitter.com/search.json?q="+
        searchTerm+"&rpp=5",
        dataType: "jsonp",
        jsonpCallback: "storeResults"
    });

}
```

20. Once we have obtained a response from the request, we'll check to make sure that we have access to the `localStorage` functionality, and if we do, we'll save the entire response, converting the JSON data into a string before we output the results to the user:

```
function storeResults(data) {
    /* Save the latest search results,
    coercing the data from an object into a string */
    localStorage.setItem(
            'twitSearchResults',
        JSON.stringify(data));
    outputResults(data);
}
```

21. We have the raw JSON data with which to create our output. Here, we loop through the results to create the required HTML blocks for displaying. As Twitter information contains a lot of links to users and dates, we'll also ensure that we have those converted for our users.

22. Within the loop, we'll append each processed result to the `tweetResults ul` tag block as an individual list item element.

23. Once the processing is complete, we call the need to refresh the list to reload the contents ready for display:

```
function outputResults(data) {

    // Clear the content and hide the results element
    $('#tweetResults').html('');

    // Loop through the results in the JSON object
    $.each(data.results,
        function(i, tweet) {
        /* Replace and define any URLs
        for inclusion in the output */
        tweet.text =
tweet.text.replace(/((https?|s?ftp|ssh)\:\/\/[^"\s\<\>]*[^.
,;'">\:\s\<\>\)\]\!])/g,
            function(url) {
                return '<a href="'+url+'">'+url+'</a>';
            }).replace(/\B@([_a-z0-9]+)/ig,
            function(reply) {
                return  reply.charAt(0)+'<a href="http://twitter.
com/'+reply.substring(1)+'">'+reply.substring(1)+'</a>';
        });

        $('#tweetResults').append('<li><img src="' +
tweet.profile_image_url + '" /><h3>@' + tweet.from_user_name + '</
h3><p>' + tweet.text + '</p><p
class="ui-li-aside"></p></li>');
    });
    // Refresh the list view
    $('#tweetResults').listview("refresh");
}
```

24. Let's now create the functions needed to handle the visual display of our button to clear the `localStorage` cache. We can reference the button `id` attribute and then apply the `css()` jQuery method to alter its style:

```
function showClearButton() {
    $("#clear_btn").css('display', 'block');
}

function hideClearButton() {
    $("#clear_btn").css('display', 'none');
}
```

> Here, we are directly changing the styles of the elements. You can amend this code to add and remove a CSS class to handle the display instead.

25. Finally, let's include our `onError` function, which will be fired if we encounter any issues along the way:

```
function onError(error) {
    alert("Error: " + JSON.stringify(error));
}
```

26. For any users running the application for the first time or with an empty cache, the application will look something like this:

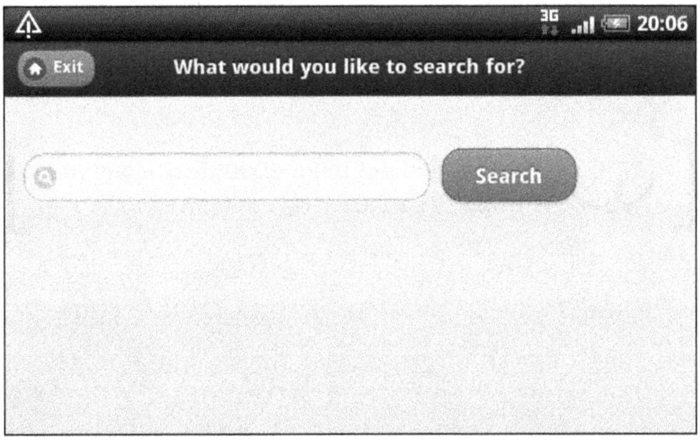

27. Once a request has been made, every time the user opens the application, they will be presented with the details of their previous search request, as shown in the following screenshot, unless they clear the cache or make a new request:

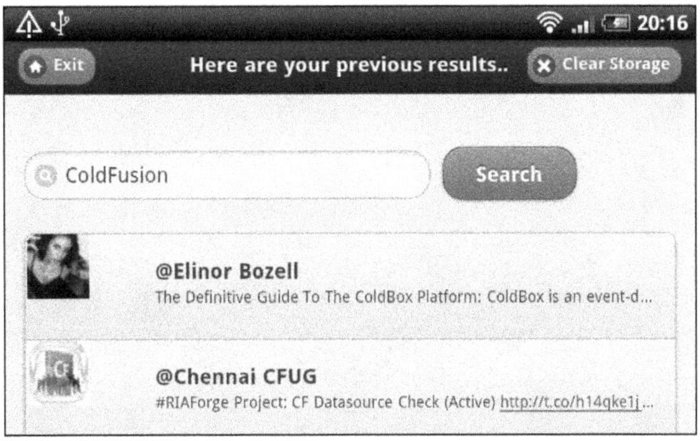

How it works...

To cache and store our data from the request, we simply saved the values, assigning them to a key that we could reference and using the `setItem` method in the `localStorage` object. We were then able to reference the storage to see whether that particular key existed by calling the `getItem` method. If nothing was present, we would make a new remote call and then save the data that was returned.

Lawnchair

The `localStorage` options provided with the Cordova API do a fantastic job of persisting and allowing us to easily access and retrieve saved data.

For those of you who wish to explore the alternative storage options, check out **Lawnchair**, an open source project written by Brian Leroux. Built with mobile applications in mind, Lawnchair is a lightweight JavaScript file that is extensible and can be used with a number of adaptors to cause data to persist, using key-value pairs, and it has an incredibly simple API.

> You can find out more about Lawnchair at `http://brian.io/lawnchair/`.

4

Working with Audio, Images, and Video

In this chapter, we will cover the following recipes:

- Capturing audio using the device audio recording application
- Recording audio within your application
- Playing audio files from the local filesystem or over HTTP
- Capturing a video using the device video recording application
- Loading a photograph from the device camera roll/library
- Applying an effect to an image using canvas
- Playing a remote video

Introduction

This chapter will include a number of recipes that outline the functionality required to capture audio, video, and camera data, as well as the playback of audio files from the local system and remote host. We will also take a look at how to use the HTML5 canvas element to edit an image on the fly.

Capturing audio using the device audio recording application

PhoneGap, through the Cordova `media-capture` plugin API, gives developers the ability to interact with the audio recording application on their device and save the recorded audio file for later use.

How to do it...

We'll make use of the `Capture` object and the `captureAudio` method of the media plugin API. The method will invoke the native device audio recording application to record our audio:

1. Firstly, create a new PhoneGap project named `audiorecord` by running the following command:

    ```
    phonegap create audiorecord com.myapp.audiorecord audiorecord
    ```

2. Add the device's platform. You can choose to use Android, iOS, or both:

    ```
    cordova platform add ios
    cordova platform add android
    ```

3. Add the `media-capture` plugin by running the following command:

    ```
    cordova plugin add org.apache.cordova.media-capture
    ```

4. Open `www/index.html`. Let's clean up the unnecessary elements. We will use jQuery Mobile, so we have to make a reference. We'll also set a style sheet reference pointing to `style.css`:

    ```
    <!DOCTYPE html>
    <html>
        <head>
            <meta charset="utf-8" />
            <meta name="format-detection"
    content="telephone=no" />
            <meta name="msapplication-tap-highlight"
    content="no" />
            <!-- WARNING: for iOS 7, remove the width=device-width and
    height=device-height attributes. See
    https://issues.apache.org/jira/browse/CB-4323 -->
            <meta name="viewport" content="user-scalable=no,
    initial-scale=1, maximum-scale=1, minimum-scale=1,
    width=device-width, height=device-height, target-
    densitydpi=device-dpi" />
            <link rel="stylesheet" type="text/css"
    href="style.css" />
            <script src="jquery/jquery-1.8.0.min.js"></script>
            <title>Hello World</title>
        </head>
        <body>

            <script type="text/javascript"
    src="cordova.js"></script>
            <script type="text/javascript">
    ```

```
        </script>
    </body>
</html>
```

5. Create a new `button` element within the `body` tags of the document, and set the `id` attribute to `record`. We'll use this to bind a touch handler to it:

```
<button id="record">capture audio</button>
```

6. Create a new file called `style.css`, and include some CSS to format the `button` element:

```
#record {
  display: block;
  padding: .4em .8em;
  text-decoration: none;
  text-shadow: 1px 1px 1px rgba(0,0,0,.3);
  -webkit-transition:.3s -webkit-box-shadow, .3s padding;
  transition:.3s box-shadow, .3s padding;
  border-radius: 200px;
  background: rgba(255,0,0,.6);
  width: 10em;
  height: 10em;
  color: white;
  position: absolute;
  top: 25%;
  left: 25%;
}
```

With the user interface added to the page and the styles applied, the application looks something like this:

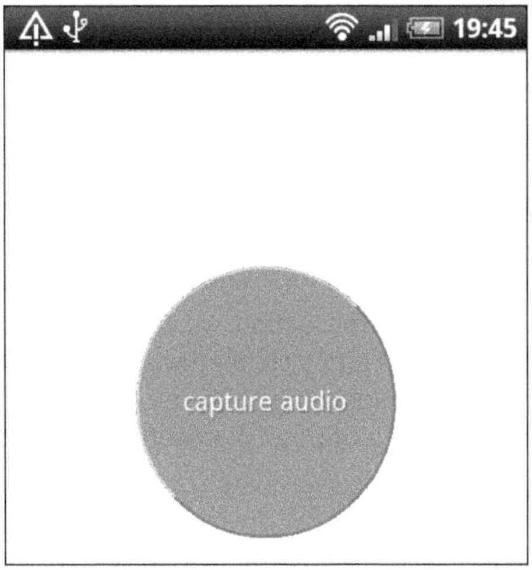

7. Now, let's start adding our custom code. Create a new `script` tag block before the closing `head` tag. Within this, we'll set up our event listener, which will call the `onDeviceReady` method once the PhoneGap code is ready to run.
8. We'll also create a global variable called `audioCapture`. It will hold our `capture` object:

```
<script type="text/javascript">
    document.addEventListener("deviceready",onDeviceReady, true);

    var audioCapture = '';
</script>
```

9. We now need to create the `onDeviceReady` method. It will assign the `capture` object to the variable that we defined earlier. We'll also bind a `touchstart` event to the `button` element, which when pressed will run the `getAudio` method to commence the capture process:

```
function onDeviceReady() {
    audioCapture = navigator.device.capture;

    $('#record').bind('touchstart', function() {
        getAudio();
    });
}
```

10. To begin the audio capture, we need to call the `captureAudio()` method from the global `capture` object. This function accepts three parameters. The first is the name of the method to run after a successful transaction. The second is the name of the error handler method to run if we encounter any problems trying to obtain the audio. The third is an array of configuration options for the capture request.
11. In this example, we are forcing the application to retrieve only one audio capture, which is also the default value:

```
function getAudio() {
    audioCapture.captureAudio(
        onSuccess,
        onError,
        {limit: 1}
    );
}
```

12. Continuing from a successful transaction, we will receive an array of objects containing the details of each audio file that was captured. We'll loop over this array and generate a string containing all the properties for each file, which we'll insert into the DOM before the `button` element:

```
function onSuccess(audioObject) {
    var i, output = '';
```

```
    for (i = 0; i < audioObject.length; i++) {
        output += 'Name: ' + audioObject[i].name + '<br />'
+
            'Full Path: ' + audioObject[i].fullPath + '<br
/>' +
            'Type: ' + audioObject[i].type + '<br />' +
            'Created: '
    + new Date(audioObject[i].lastModifiedDate) + '<br />'
+
             'Size: ' + audioObject[i].size + '<br
/>========';
    }

    $('#record').before(output);
}
```

13. If we encounter an error during the process, the `onError` method will fire. This method will provide us with access to an error object, that contains the code for the error. We can use a `switch` statement here to customize the message that we will return to our user:

```
function onError(error) {
    var errReason;
    switch(error.code) {
        case 0:
            errReason = 'The microphone failed to capture
sound.';
        break;
       case 1:
           errReason = 'The audio capture application is
currently busy with another request.';
       break;
       case 2:
           errReason = 'An invalid parameter was sent to
the API.';
       break;
       case 3:
           errReason = 'You left the audio capture
application without recording anything.';
       break;
       case 4:
           errReason = 'Your device does not support the
audio capture request.';
       break;
    }
    alert('The following error occurred: ' + errReason);
}
```

If we run our application and press the button, the device's default audio recording application will open and we will be able to record our audio, as follows:

14. Once we have finished recording, our application will receive the audio data from the callback method and output it like this:

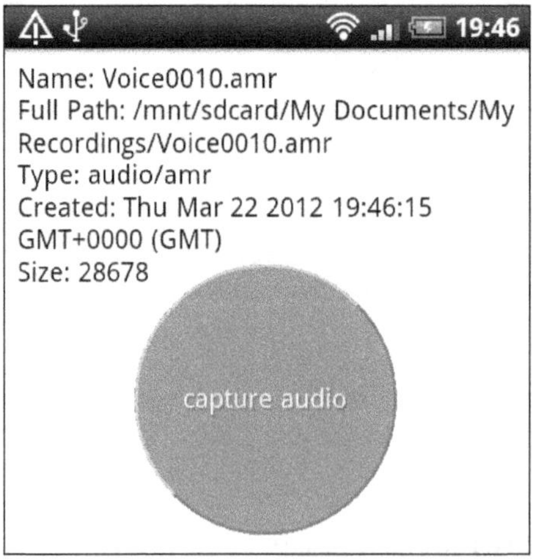

How it works...

The `Capture` object, available through the `media-capture` API, allows us to access the `media-capture` capabilities of the device. By specifying the media type that we wish to capture by calling the `captureAudio` method, an asynchronous call is made to the device's native audio recording application.

In this example, we requested the capture of only one audio file. Setting the limit value within the optional configuration to a value greater than 1 can alter this.

The request is completed when one of these two things happens:

- The maximum number of recordings possible have been created
- The user exits the native audio recording application

Following a successful callback from the request operation, we receive an array of objects. This array contains the properties for each individual media file, which contains the following properties that we can read:

- `name`: A `DOMString` that contains the name of the file
- `fullPath`: A `DOMString` that contains the full path of the file
- `type`: A `DOMString` that includes the MIME type of the returned media file
- `lastModifiedTime`: A `Date` object that contains the date and time when the file was last modified
- `size`: A `Number` that contains the size of the file in bytes

> To find out more about the `captureAudio` capabilities offered by the PhoneGap API, check out the official documentation at `http://plugins.cordova.io/#/package/org.apache.cordova.media-capture`.

See also

- The *Playing audio files from the local filesystem or over HTTP* section of this chapter

Recording audio within your application

The PhoneGap API provides us with the ability to record audio directly within our application, bypassing the native audio recording application.

How to do it...

We will use the `Media` object to create a reference to an audio file in which we'll record the audio data:

1. Firstly, create a new PhoneGap project named `audioapi` by running this line:

   phonegap create audioapi com.myapp.audioapi audioapi

2. Add the device's platform. You can choose to use Android, iOS, or both:

   cordova platform add ios

   cordova platform add android

3. Add the `media-capture` plugins by running this line:

   cordova plugin add org.apache.cordova.media

4. Open `www/index.html`. Let's clean up the unnecessary elements. We will use jQuery Mobile, so we have to make a reference:

```
<!DOCTYPE html>
<html>
    <head>
        <meta charset="utf-8" />
        <meta name="format-detection"
content="telephone=no" />
        <meta name="msapplication-tap-highlight"
content="no" />
        <meta name="viewport" content="user-scalable=no,
initial-scale=1, maximum-scale=1, minimum-scale=1,
width=device-width, height=device-height, target-
densitydpi=device-dpi" />
        <link type="text/css"
href="jquery/css/smoothness/jquery-ui-1.8.23.custom.css"
rel="stylesheet" />
        <script type="text/javascript" src="jquery/jquery-
1.8.0.min.js"></script>
        <script type="text/javascript" src="jquery/jquery-
ui-1.8.23.custom.min.js"></script>
        <title>Hello World</title>
    </head>
    <body>
```

```
        <script type="text/javascript"
src="cordova.js"></script>
        <script type="text/javascript">

        </script>
    </body>
</html>
```

5. We include three elements within the `body` of our application. The first is a `div` with the `id` attribute set to `progressbar`, the second is a `div` tag with the `id` attribute set to `message`, and the third is a `button` element with the `id` attribute set to `record`:

```
<div id="progressbar"></div>
<div id="message"></div>
<button id="record"></button>
```

6. Now, let's start adding our custom code within the empty `script` tag block. We'll begin by defining some global variables that we'll use in the application. We'll also create the event listener to ensure that the device is ready before we proceed.
7. Then, the `onDeviceReady` function will run a new function called `recordPrepare`:

```
var maxTime = 10,
    countdownInt = 3,
    src,
    audioRecording,
    stopRecording;

document.addEventListener("deviceready",  onDeviceReady,
false);

function onDeviceReady() {
    recordPrepare();
}
```

8. The `recordPrepare` button will be used more than once in our application to reset the state of the button to record audio. Here, we unbind any actions applied to the button, set the HTML value, and bind the `touchstart` handler to run a function called `recordAudio`:

```
function recordPrepare() {
    $('#record').unbind();
    $('#record').html('Start recording');
    $('#record').bind('touchstart', function() {
        recordAudio();
    });
}
```

9. Let's now create the `recordAudio()` function, which will create the audio file. We'll switch the value and bind events applied to our button to allow the user to manually end the recording. We will also set the `Media` object to a variable, `audioRecording`, and pass the destination for the file in the form of the `src` parameter as well as the success and error callback methods.

10. A `setInterval` method is included. It will count down from 3 to 0 to give the user some time to prepare for the recording. When the countdown is complete, we invoke the `startRecord` method from the `Media` object and start another `setInterval` method. This will count to 10 and automatically stop the recording when the limit is reached:

```
function recordAudio() {

    $('#record').unbind();
    $('#record').html('Stop recording');
    $('#record').bind('touchstart', function() {
        stopRecording();
    });

    src = 'recording_' + Math.round(new
Date().getTime()/1000) + '.mp3';

    audioRecording = new Media(src, onSuccess, onError);

    var startCountdown = setInterval(function() {

        $('#message').html('Recording will start in ' +
countdownInt + ' seconds...');
        countdownInt = countdownInt -1;

        if(countdownInt <= 0) {
            countdownInt = 3;
            clearInterval(startCountdown);
            audioRecording.startRecord();

            var recTime = 0;
                recInterval = setInterval(function() {
                recTime = recTime + 1;

                $('#message').html(Math.round(maxTime -
recTime) + ' seconds remaining...');
```

```
                var progPerc = 100-((100/maxTime) *
recTime);
                setProgress(progPerc);

                if (recTime >= maxTime) {
                    stopRecording();
                }
            }, 1000);
        }
    }, 1000);
}
```

11. While our recording is underway, we can update the progress bar using the jQuery UI library and set it to the current value to show how much time is remaining:

```
function setProgress(progress) {
    $("#progressbar").progressbar({
            value: progress
    });
}
```

12. When a recording is stopped, we want to clear the interval timer and run the `stopRecord` method from the `Media` object. We'll also reset the value of the progress bar to 0 and reset the button bindings to prepare for the next recording:

```
function stopRecording() {
    clearInterval(recInterval);
    audioRecording.stopRecord();
    setProgress(0);
    recordPrepare();
}
```

13. Finally, we can add in our success and error callback methods:

```
function onSuccess() {
    $('#message').html('Audio file successfully
    created:<br />' + src);
}

function onError(error) {
    $('#message').html('code: ' + error.code    + '\n' +
          'message: ' + error.message + '\n');
}
```

14. When the application is run to start the recording, the output will look somewhat like this:

15. After a successful recording, the user will be presented with the URI to the recorded file, which we can use to access, upload, or play back the file.

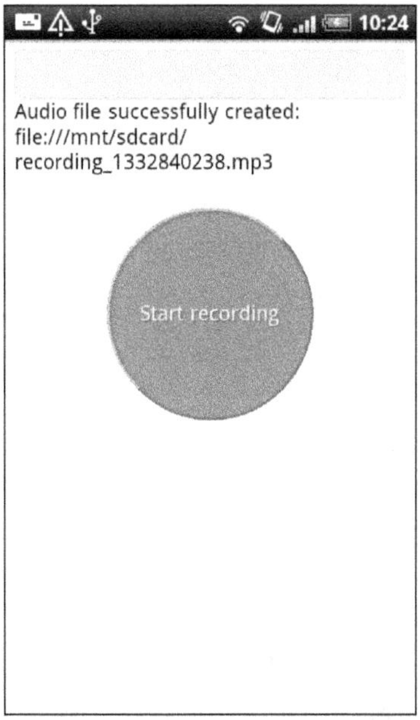

How it works...

The `Media` object has the ability to record and play back audio files. When we choose to use this object for recording, we need to provide the method with the URI for the destination file on the local device.

To start a recording, we simply call the `startRecord` method from the `Media` object, and to stop the recording, we need to call the `stopRecord` method.

> To find out more about the available methods within the `Media` object, refer to the official documentation, available at `http://plugins.cordova.io/#/package/org.apache.cordova.media`.

See also

- The *Saving a file in the device storage* section of *Chapter 3, Filesystems, Storage, and Local Databases*
- The *Opening a local file from the device storage* section of *Chapter 3, Filesystems, Storage, and Local Databases*

Playing audio files from the local filesystem or over HTTP

PhoneGap and the media plugin API provide us with a relatively straightforward process to play back audio files. These can be files stored within the application's local filesystem, bundled with the application, or over remote files accessible by a network connection. Wherever the files may be, the method of playback is achieved in exactly the same way.

How to do it...

We must create a new `Media` object and pass to it the location of the audio file we want to play back:

1. Firstly, create a new PhoneGap project named `audioplaying` by running the following command:

   ```
   phonegap create audioplaying com.myapp.audioplaying audioplaying
   ```

2. Add the device's platform. You can choose to use Android, iOS, or both:

   ```
   cordova platform add ios
   cordova platform add android
   ```

3. Add the `media` and `file-transfer` plugins by running the following commands:

   cordova plugin add org.apache.cordova.media

   cordova plugin add org.apache.cordova.file-transfer

4. Open `www/index.html` and clean up the unnecessary elements. We will use jQuery Mobile, so we need to make a reference:

```
<!DOCTYPE html>
<html>
    <head>
        <meta charset="utf-8" />
        <meta name="format-detection"
content="telephone=no" />
        <meta name="msapplication-tap-highlight"
content="no" />
        <meta name="viewport" content="user-scalable=no,
initial-scale=1, maximum-scale=1, minimum-scale=1,
width=device-width, height=device-height, target-
densitydpi=device-dpi" />
        <link rel="stylesheet" href="jquery/jquery.mobile-
1.1.1.min.css" type="text/css">
        <script type="text/javascript" src="jquery/jquery-
1.8.0.min.js"></script>
        <script type="text/javascript"
src="jquery/jquery.mobile-1.1.1.min.js"></script>
        <title>Hello World</title>
    </head>
    <body>

        <script type="text/javascript"
src="cordova.js"></script>
        <script type="text/javascript">
        </script>
    </body>
</html>
```

5. Create the layout for the application within the `body` tags. Here, we are specifying a page for the jQuery Mobile framework and four key `div` elements that have been assigned the role of buttons. We will reference their `id` attributes in our code:

```
<div data-role="page" id="page-home">
    <div data-role="header">
        <h1>PhoneGap Audio Player</h1>
    </div>

    <div data-role="content">
```

```
        <div data-role="button"
        id="playLocalAudio">Play Local Audio</div>
        <div data-role="button"
        id="playRemoteAudio">Play Remote Audio</div>
        <div data-role="button"
        id="pauseaudio">Pause Audio</div>
        <div data-role="button"
        id="stopaudio">Stop Audio</div>

        <div class="ui-grid-a">
            <div class="ui-block-a"> Current:
         <span id="audioPosition">0 sec</span></div>
            <div class="ui-block-b">Total:
         <span id="mediaDuration">0</span> sec</div>
        </div>

    </div>
</div>
```

6. We create a new `script` tag block within the `head` tag to contain our custom code, into which we'll add our event listener in order to check whether the device is ready to proceed, and some required global variables:

```
<script type="text/javascript">
    document.addEventListener("deviceready", onDeviceReady,
true);

    var audioMedia = null,
        audioTimer = null,
        duration = -1,
        is_paused = false;
</script>
```

7. The `onDeviceReady` method binds a `touchstart` event to all four of our buttons in the main page content. For the local audio option, this example is set to read a file from the Android asset location. In both the `play` functions, we pass the audio source to the `playAudio` method:

```
function onDeviceReady() {

    $("#playLocalAudio").bind('touchstart', function() {

        stopAudio();
        var srcLocal =
'/android_asset/www/CFHour_Intro.mp3';
        playAudio(srcLocal);
```

```
    });

    $("#playRemoteAudio").bind('touchstart', function() {

        stopAudio();
        var srcRemote =
'http://traffic.libsyn.com/cfhour/Show_138_-_ESAPI_StackOverflow_
and_Community.mp3';
        playAudio(srcRemote);

    });

    $("#pauseaudio").bind('touchstart', function() {
        pauseAudio();
    });

    $("#stopaudio").bind('touchstart', function() {
        stopAudio();
    });

}
```

8. Now, let's add the `playAudio` method. This will check whether the `audioMedia` object has been assigned and we have an active audio file. If not, we will reset the duration and position values and create a new `Media` object reference, passing in the source of the audio file.
9. To update the duration and current position of the audio file, we will set a new interval timer. It will check once every second and obtain these details from the `getCurrentPosition` and `getDuration` methods, available from the `Media` object:

```
function playAudio(src) {

    if (audioMedia === null) {
        $("#mediaDuration").html("0");
        $("#audioPosition").html("Loading...");
        audioMedia = new Media(src, onSuccess, onError);
        audioMedia.play();
    } else {
        if (is_paused) {
            is_paused = false;
            audioMedia.play();
        }
    }
```

```
    if (audioTimer === null) {
        audioTimer = setInterval(function() {
audioMedia.getCurrentPosition(
function(position) {
  if (position > -1) {

  setAudioPosition(Math.round(position));
    if (duration <= 0) {
   duration = audioMedia.getDuration();
        if (duration > 0) {
          duration = Math.round(duration);
       $("#mediaDuration").html(duration);
        }
    }
  }
},
function(error) {
  console.log("Error getting position=" + error);
  setAudioPosition("Error: " + error);
}
        );
        }, 1000);
    }
}
```

10. The `setAudioPosition` method will update the content in the `audioPosition` element with the current details:

```
function setAudioPosition(position) {
    $("#audioPosition").html(position + " sec");
}
```

11. Now we can include the two remaining methods assigned to the touch handlers to control the pausing and stopping of the audio playback:

```
function pauseAudio() {
    if (is_paused) return;
    if (audioMedia) {
        is_paused = true;
        audioMedia.pause();
    }
}

function stopAudio() {
    if (audioMedia) {
        audioMedia.stop();
```

```
            audioMedia.release();
            audioMedia = null;
        }
        if (audioTimer) {
            clearInterval(audioTimer);
            audioTimer = null;
        }

        is_paused = false;
        duration = 0;
    }
```

12. Finally, let's write the success and error callback methods. In essence, they both reset the values to the default positions in preparation for the next playback request:

```
    function onSuccess() {
        setAudioPosition(duration);
        clearInterval(audioTimer);
        audioTimer = null;
        audioMedia = null;
        is_paused = false;
        duration = -1;
    }

    function onError(error) {
        alert('code: ' + error.code + '\n' +
                'message: ' + error.message + '\n');
        clearInterval(audioTimer);
        audioTimer = null;
        audioMedia = null;
        is_paused = false;
        setAudioPosition("0");
    }
```

13. Run the application on the device. The output will be similar to what is shown here:

How it works...

The `Media` object has the ability to record and play back audio files. For media playback, we simply pass the location of the audio file, remote or local, to the `Media` instantiation call, along with the success and error handlers.

Playback is controlled by the `Media` objects' methods available through the media plugin API.

> To find out more about all the available methods within the `Media` object, refer to the official documentation at `http://plugins.cordova.io/#/package/org.apache.cordova.media`.

There's more...

In this example, we assumed that the developer is building for an Android device, and so we referenced the location of the local file using the `android_asset` reference. To cater to other devices' operating systems, you can use the `Device` object, available in the device plugin, to determine which platform is running the application. Using the response from this check, you can write a `switch` statement to provide the correct path to the local file.

> To learn more about the `Device` object, refer to the official documentation, which is available at `http://plugins.cordova.io/#/package/org.apache.cordova.device`.

Capturing a video using the device video recording application

PhoneGap, through the media plugin API, provides us with the ability to easily access the native video recording application on our mobile device and save the captured footage.

How to do it...

We will use the `Capture` object and the `captureVideo` method that it contains to invoke the native video recording application:

1. Firstly, create a new PhoneGap project named `videorecording` by running the following command:

   ```
   phonegap create videorecording com.myapp.videorecording
   videorecording
   ```

2. Add the device's platform. You can choose to use Android, iOS, or both:

   ```
   cordova platform add ios
   cordova platform add android
   ```

3. Add the `media` and `file-transfer` plugins by running the following commands:

   ```
   cordova plugin add org.apache.cordova.media
   cordova plugin add org.apache.cordova.file-transfer
   ```

4. Open `www/index.html`. Let's clean up the unnecessary elements. We will use jQuery Mobile, so we have to make a reference. We'll also set a style sheet reference pointing to `style.css`:

   ```
   <!DOCTYPE html>
   <html>
       <head>
           <meta charset="utf-8" />
           <meta name="format-detection"
   content="telephone=no" />
           <meta name="msapplication-tap-highlight"
   content="no" />

           <meta name="viewport" content="user-scalable=no,
   initial-scale=1, maximum-scale=1, minimum-scale=1,
   width=device-width, height=device-height, target-
   densitydpi=device-dpi" />
           <link rel="stylesheet" type="text/css"
   href="style.css" />
           <script type="text/javascript"
               src="jquery/jquery-1.8.0.min.js"></script>
           <title>Hello World</title>
       </head>
       <body>

           <script type="text/javascript"
   src="cordova.js"></script>
           <script type="text/javascript">
           </script>
       </body>
   </html>
   ```

5. Create a new `button` element within the `body` tags of the document, and set the `id` attribute to `record`. We'll use this to bind a touch handler to it:

   ```
   <button id="record">capture video</button>
   ```

6. Now, let's start adding our custom code. Create a new `script` tag block before the closing `head` tag. Within this, we'll set up our event listener, which will call the `onDeviceReady` method once the PhoneGap code is ready to run.
7. We'll also create a global variable called `videoCapture`. It will hold our `capture` object:

```
document.addEventListener("deviceready", onDeviceReady, true);

var videoCapture = '';
```

8. We now need to create the `onDeviceReady` method. This method will assign the `capture` object to the variable we defined earlier. We'll also bind a `touchstart` event to the `button` element, which when pressed will run the `getVideo` method to commence the `capture` process:

```
function onDeviceReady() {
   videoCapture = navigator.device.capture;

   $('#record').bind('touchstart', function() {
        getVideo();
    });
}
```

9. To begin the video capture, we need to call the `captureVideo` method from the global `capture` object. This function accepts three parameters. The first is the name of the method to run after a successful transaction. The second is the name of the error handler method to run if we encounter any problems trying to obtain the video. Finally, the third is an array of configuration options for the capture request.
10. In this example, we are requesting the application to retrieve two separate video captures:

```
function getVideo() {
    videoCapture.captureVideo(
        onSuccess,
        onError,
        {limit: 2}
    );
}
```

11. Following on from a successful transaction, we will receive an array of objects containing the details for each video file that was captured. We'll loop over this array and generate a string containing all the properties for each file, which we'll insert into the DOM before the `button` element:

```
function onSuccess(videoObject) {
    var i, output = '';
    for (i = 0; i < videoObject.length; i += 1) {
```

```
        output += 'Name: ' + videoObject[i].name + '<br />'
+
            'Full Path: ' + videoObject[i].fullPath + '<br />' +
             'Type: ' + videoObject[i].type + '<br />' +
             'Created: '
    + new Date(videoObject[i].lastModifiedDate) + '<br />'
+
             'Size: ' + videoObject[i].size + '<br
/>========';
        }
    $('#record').before(output);
}
```

12. If we encounter an error during the process, the `onError` method will fire. This method will provide us with access to an `error` object, which contains the code for the error. We can use a `switch` statement here to customize the message that we will return to our user:

```
function onError(error) {
    var errReason;
       switch(error.code) {
        case 0:
            errReason = 'The camera failed to capture
video.';
        break;
        case 1:
           errReason = 'The video capture application is
currently busy with another request.';
        break;
        case 2:
           errReason = 'An invalid parameter was sent to
the API.';
        break;
        case 3:
           errReason = 'You left the video capture
application without recording anything.';
        break;
        case 4:
           errReason = 'Your device does not support the
video capture request.';
        break;
    }
    alert('The following error occurred: ' + errReason);
}
```

13. If we run our application and press the button, the device's default video recording application will open, and we can record our video, as shown in this screenshot:

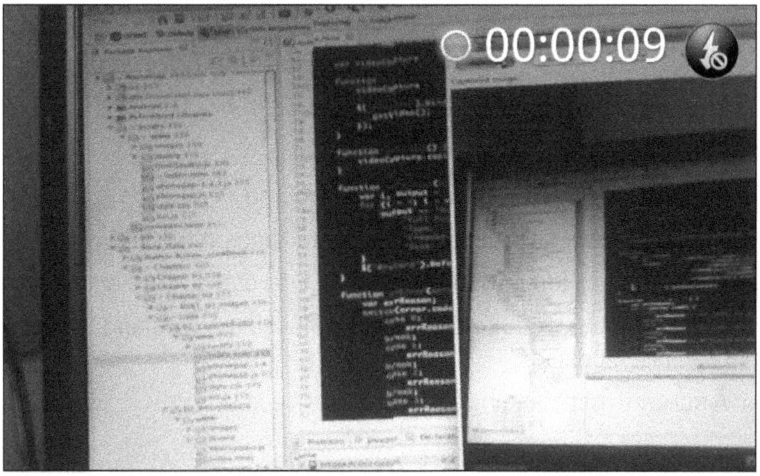

14. Once we have finished recording, our application will receive the video data from the callback method and output it, like this:

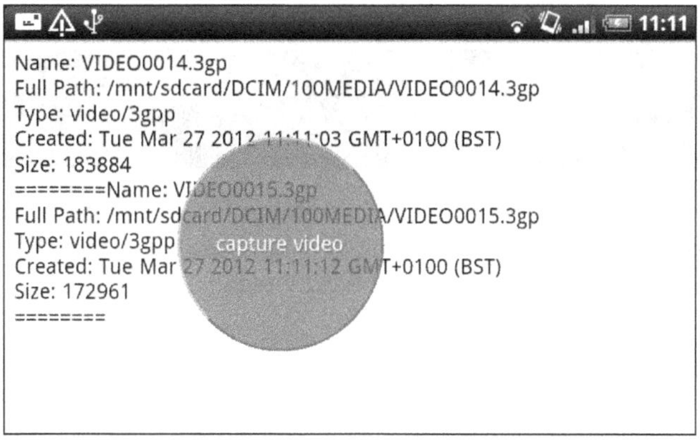

How it works...

The `Capture` object, available through the media-capture API, grants us access to the media capture capabilities of the device. By specifying the media type that we wish to capture by calling the `captureVideo` method, an asynchronous call is made to the device's native video recording application.

In this example, we forced the method to request two video captures by setting the `limit` property in the optional configuration options—the default value for this limit is set to 1.

The request is returned when one of these two things happens:

- The maximum number of recordings have been created
- The user exits the native video recording application

Following a successful callback from the request operation, we receive an array of objects. This array contains properties for each individual media file, which contains the following properties that we can read:

- `name`: A `DOMString` that contains the name of the file
- `fullPath`: A `DOMString` that contains the full path of the file
- `type`: A `DOMString` that includes the MIME type of the returned media file
- `lastModifiedTime`: A `Date` object that contains the date and time at which the file was last modified
- `size`: A `Number` that contains the size of the file in bytes

> To find out more about the `captureVideo` capabilities offered by the PhoneGap API, check out the official documentation at `http://plugins.cordova.io/#/package/org.apache.cordova.media-capture`.

Loading a photograph from the device camera roll/library

Different devices will store saved photographs in different locations, typically in either a photo library or a saved photo album. The camera plugin API gives developers the ability to select or specify from which location an image should be retrieved.

How to do it...

We must use the `getPicture` method, available from the `Camera` object, to either select an image from the device library, or capture a new image directly from the camera:

1. Firstly, create a new PhoneGap project named `cameraroll` by running this command:

   ```
   phonegap create cameraroll com.myapp.cameraroll cameraroll
   ```

2. Add the devices platform. You can choose to use Android, iOS, or both:

   ```
   cordova platform add ios
   cordova platform add android
   ```

3. Add the `media` and `file-transfer` plugins by running the following commands:

   ```
   cordova plugin add cordova-plugin-media
   cordova plugin add cordova-plugin-file-transfer
   ```

4. Open `www/index.html`. Let's clean up the unnecessary elements. We will use jQuery Mobile, so we need to make a reference:

   ```
   <!DOCTYPE html>
   <html>
       <head>
           <meta charset="utf-8" />
           <meta name="format-detection"
   content="telephone=no" />
           <meta name="msapplication-tap-highlight"
   content="no" />
           <meta name="viewport" content="user-scalable=no,
   initial-scale=1, maximum-scale=1, minimum-scale=1,
   width=device-width, height=device-height, target-
   densitydpi=device-dpi" />
           <script type="text/javascript" src="jquery/jquery-
   1.8.0.min.js"></script>
           <title>Hello World</title>
       </head>
       <body>

           <script type="text/javascript"
   src="cordova.js"></script>
           <script type="text/javascript">

           </script>
       </body>
   </html>
   ```

5. The `body` of our application will contain four elements. We'll need to provide two buttons, both with the `class` attribute set to `photo`, and each of them with the `id` attribute set to `cameraPhoto` and `libraryPhoto` respectively.
6. We also need to create a `div` element with `id` set to `message`, and an `img` tag with `id` set to `image`:

   ```
   <button class="photo" id="cameraPhoto">Take New
   Photo</button>
   <br />
   <button class="photo" id="libraryPhoto">Select From
   Library</button><br />
   <div id="message"></div><br />
   <img id="image" />
   ```

7. Create a new `script` tag block within the head of the document, and include the event listener that will fire when the PhoneGap native code is compiled and ready. Below this, we will create the `onDeviceReady` function, within which we will apply a bind handler to the buttons using the jQuery class selector.

8. Depending on the value of the selected button's `id` attribute, the `switch` statement will run the particular method to obtain the image:

```
function onDeviceReady() {
    $('.photo').bind('touchstart', function() {
        switch ($(this).attr('id')) {
            case 'cameraPhoto':
                capturePhoto();
                break;
            case 'libraryPhoto':
                getPhoto();
                break;
        }
    });
}
```

9. Let's now add the first of our image capture functions—`capturePhoto`. This function calls the `getPicture` method from the `Camera` object. Here, we are asking for the highest quality image returned and a scaled image to match the provided sizes:

```
function capturePhoto() {
    navigator.camera.getPicture(onSuccess, onFail, {
        quality: 100,
        targetWidth: 250,
        targetHeight: 250
    });
}
```

10. The second image capture method is `getPhoto`. In this method, we call the `getPicture` method once again, but this time, we pass the `sourceType` option value to request that the image be selected from the device photo library:

```
function getPhoto() {
    navigator.camera.getPicture(onSuccess, onFail, {
        quality: 100,
        destinationType: Camera.DestinationType.FILE_URI,
        sourceType: Camera.PictureSourceType.PHOTOLIBRARY,
        targetWidth: 250,
        targetHeight: 250
    });
}
```

11. Finally, let's add the success and error handlers, which will be used by both of our capture methods. The `onSuccess` method will display the returned image, setting it as the source for the `image` element:

```
function onSuccess(imageURI) {
    $('#image').attr('src', imageURI);
    $('#message').html('Image location: ' + imageURI);
}

function onFail(message) {
    $('#message').html(message);
}
```

12. Upon running the application on the device, the output will look something like what is shown here:

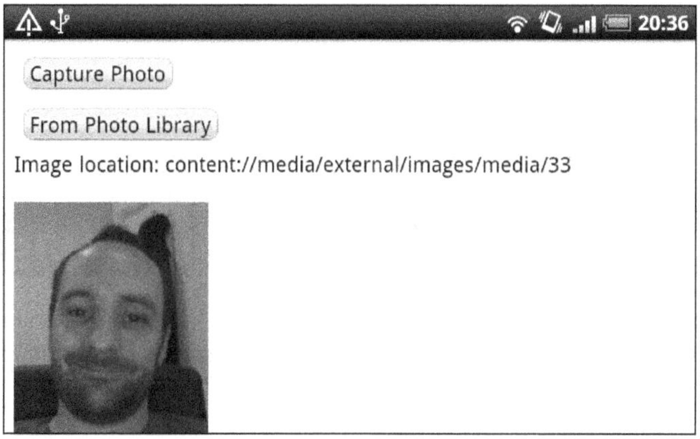

How it works...

The `Camera` object, available through the camera plugin API, allows us to interact with the default camera application on the device. The `Camera` object itself contains only one method—`getPicture`. Depending on the `sourceType` value being sent through the `capture` request method, we can obtain the image either from the device camera or by selecting a saved image from the photo library or photo album.

In this example, we retrieved the URI for the image to use it as the source for an `img` tag. This method can also return the image as a Base64-encoded image, if requested.

> There are a number of optional parameters that we can send to the method calls to customize the camera settings. For detailed information about each parameter, refer to the official documentation available at `http://plugins.cordova.io/#/package/org.apache.cordova.camera`.

There's more...

In this recipe, we requested that the images be scaled to match a certain dimension, while maintaining the aspect ratio. When selecting an image from the library or saved album, PhoneGap resizes the image and stores it in a temporary cache directory on the device. While this means that resizing is as painless as we would want it to be, the image may not persist, or it will be overwritten when the next image is resized.

If you want to save resized images in a permanent location after creating them, make sure that you check out the recipes within this book on how to interact with the local filesystem and how to save files.

Now that we can easily obtain an image from the device, there are a number of things we can do with it. For an example, take a look at the next recipe in this chapter.

See also

- The *Uploading a file on a remote server* recipe in *Chapter 3, Filesystems, Storage, and Local Databases*

Applying an effect to an image using canvas

Capturing a photo on our device is fantastic, but what all can we do with an image once we have it in our application? In this recipe, we'll create some simple functions to edit the color of an image without altering the original source.

How to do it...

We must create and use the HTML5 canvas element to load and edit the values of a stored image, by performing the following steps:

1. Firstly, create a new PhoneGap project named `cameraeffect` by running the following command:

   ```
   phonegap create cameraeffect com.myapp.cameraeffect cameraeffect
   ```

2. Add the devices platform. You can choose to use Android, iOS, or both:

   ```
   cordova platform add ios
   cordova platform add android
   ```

3. Open `www/index.html` and clean up the unnecessary elements. We will use jQuery, so we have to make a reference:

```
<!DOCTYPE html>
<html>
    <head>
        <meta charset="utf-8" />
        <meta name="format-detection"
content="telephone=no" />
        <meta name="msapplication-tap-highlight"
content="no" />
        <meta name="viewport" content="user-scalable=no,
initial-scale=1, maximum-scale=1, minimum-scale=1,
width=device-width, height=device-height, target-
densitydpi=device-dpi" />
        <script type="text/javascript" src="jquery/jquery-
1.8.0.min.js"></script>
        <title>Hello World</title>
    </head>
    <body>

        <script type="text/javascript"
src="cordova.js"></script>
        <script type="text/javascript">

        </script>
    </body>
</html>
```

4. Include a reference to the `rgb.js` file that is available in the project download that accompanies this book, below the Cordova JavaScript reference. This contains the required array of variables for one of our image manipulation functions:

```
<script type="text/javascript" src="cordova.js"></script>
<script type="text/javascript" src="rgb.js"></script>
<script type="text/javascript">
```

5. The `body` of our application will hold three `button` elements, each with a specific `id` attribute that we will reference within the custom code. We'll also need an `img` tag with the `id` attribute set to `sourceImage`, which will display the original image that we want to manipulate.
6. Finally, we need to include a `canvas` element with the `id` attribute set to `myCanvas`:

```
<button id="grayscale">Grayscale</button>
<button id="sepia">Sepia</button>
<button id="reset">Reset</button><br />

<img id="sourceImage" src="awesome.jpg" alt="source image"
height="150" width="150" />
```

7. Let's start adding our custom code. We will use the `script` tag block before the closing `body` tag, into which we'll add our event listener to ensure that PhoneGap is fully loaded before we proceed. We'll also create some required global variables:

```
document.addEventListener("deviceready", onDeviceReady, true);

var canvas,
        context,
        image,
        imgObj,
        noise = 20;
```

8. Create the `onDeviceReady` method, which will run once the native code is ready. Here, we want to run a method called `reset`. It will restore our `canvas` to its default source. We'll also bind the `touchstart` handlers to our three buttons, each of which will run its own method:

```
function onDeviceReady() {
    reset();

    $('#grayscale').bind('touchstart', function() {
        grayscaleImage();
    });

    $('#sepia').bind('touchstart', function() {
        processSepia();
    });

    $('#reset').bind('touchstart', function() {
        reset();
    });
}
```

9. The `reset` method creates the `canvas` reference and its context, and applies the source from our starting image to it:

```
function reset() {
    canvas = document.getElementById('myCanvas');
    context = canvas.getContext("2d");
    image = document.getElementById('sourceImage');
    context.drawImage(image, 0, 0);
}
```

10. Our first image manipulation function is called `grayscaleImage`. Let's include it now. Within it, we'll loop through the pixel data of our image, which we can retrieve from the `canvas` element using the `getImageData` method:

```
function grayscaleImage() {
    var imageData = context.getImageData(0, 0, 300, 300);
    for (var i = 0, n = imageData.data.length; i < n; i +=
4) {
        var grayscale = imageData.data[i] * .3 +
            imageData.data[i + 1] * .59 + imageData.data[i
+ 2] * .11;
        imageData.data[i] = grayscale;
        imageData.data[i + 1] = grayscale;
        imageData.data[i + 2] = grayscale;
    }
    context.putImageData(imageData, 0, 0);
}
```

11. Our second manipulation function is called `processSepia`. Once again, we will obtain the image data from our `canvas` element and loop through each pixel, applying the changes as we go:

```
function processSepia() {
    var imageData =
        context.getImageData(0, 0, canvas.width,
canvas.height);
    for (var i = 0; i < imageData.data.length; i += 4) {
        imageData.data[i] = r[imageData.data[i]];
        imageData.data[i + 1] = g[imageData.data[i + 1]];
        imageData.data[i + 2] = b[imageData.data[i + 2]];
        if (noise > 0) {
            var noise = Math.round(noise - Math.random() *
noise);
            for (var j = 0; j < 3; j++) {
                var iPN = noise + imageData.data[i + j];
                imageData.data[i + j] = (iPN > 255) ? 255 :
iPN;
            }
        }
    }
    context.putImageData(imageData, 0, 0);
};
```

12. Upon running the application on the device after selecting a button to change our default image, the output will look something like what is shown in the following screenshot:

How it works...

When we start processing a change to the canvas image, we first obtain the data using the `getImageData` method, which is available through the `canvas` context. We can easily access the information for each pixel within the returned image object and its `data` attribute.

With the data in an array, we can loop over each pixel object first, and then over each value within each pixel object.

> Pixels contain four values: the red, green, blue, and alpha channels.

By looping over each color channel in each pixel, we can alter the values, thereby changing the image. We can then set the revised image as the source in our canvas, using the `putImageData` method to set it back to the context of our `canvas`.

There's more...

Although this recipe does not involve any PhoneGap-specific code with the exception of the `onDeviceReady` method, it was included here for three reasons:

- As an example to show you how you might like to work with images captured using the camera plugin API
- To remind you of or introduce you to the power of HTML5 elements, and demonstrate how you can work with the canvas
- Because it's pretty cool!

Playing a remote video

In reach media mobile applications, we are often required to play videos from a remote source. PhoneGap can play remote videos by two ways: a plugin or a web API. The simpler way is by using a web API.

How to do it...

Playing a video using PhoneGap can be done by using the HTML5 `video` tag:

1. Firstly, create a new PhoneGap project named `remotevideo` by running this command:

   phonegap create remotevideo com.myapp.remotevideo remotevideo

2. Add the devices platform. You can choose to use Android, iOS, or both:

   cordova platform add ios

   cordova platform add android

3. Open `www/index.html`. Let's clean up the unnecessary elements. So, we have the following code:

```
<!DOCTYPE html>
<html>
    <head>
        <meta charset="utf-8" />
        <meta name="format-detection"
content="telephone=no" />
        <meta name="msapplication-tap-highlight"
content="no" />
        <meta name="viewport" content="user-scalable=no,
initial-scale=1, maximum-scale=1, minimum-scale=1,
width=device-width, height=device-height, target-
densitydpi=device-dpi" />
```

```
        <title>Hello World</title>
    </head>
    <body>

        <script type="text/javascript"
src="cordova.js"></script>
        <script type="text/javascript">

        </script>
    </body>
</html>
```

4. Let's add the `video` tag that holds the URL of the video. We will be using a sample video from `http://w3schools.com`:

```
<div align="center">
    <video width="320" height="240" controls>
        <source
src="http://www.w3schools.com/tags/movie.mp4"
type="video/mp4">
        <source
src="http://www.w3schools.com/tags/movie.ogg"
type="video/ogg">
    </video>
</div>
```

5. Running the application on the device will give us the following result:

How it works...

The `video` tag on a PhoneGap application works just as on a web application. We add a reference to the video URL using the `source` tag. We can also use other `video` attributes such as `control` to display or hide the video control.

5
Working with Your Contacts List

In this chapter, we will cover these recipes:

- Listing all available contacts
- Displaying the contact information of a specific individual
- Creating and saving a new contact

Introduction

With ever-expanding and increasing technological resources and advancements, mobile devices contain increasingly powerful processors and provide users and consumers with an impressive array of features.

Above all the apps, widgets, and features that your device can manage, let's not forget that the primary function of a mobile device (certainly a mobile phone) is to hold the contact information of your friends, family, or favorite local takeaway restaurants.

All the recipes in this chapter will focus on interacting with your device's contact database and how you can list, display, and add contacts into it.

Listing all available contacts

Developers can access, read from, and filter the contacts saved within the device's contact database, allowing them to query and work with the address book on the device.

How to do it...

We will create an application to read all contacts in the device database and output them as a list:

1. Firstly, create a new PhoneGap project named `readcontact` by running the following command:

   ```
   phonegap create readcontact com.myapp.readcontact readcontact
   ```

2. Add the devices platform. You can choose to use Android, iOS, or both:

   ```
   cordova platform add ios
   cordova platform add android
   ```

3. Add the contact plugins by running this command:

   ```
   cordova plugin add org.apache.cordova.contacts
   ```

4. Open `www/index.html`, and let's clean up the unnecessary elements. We will be using jQuery Mobile, so we have to make a reference:

   ```
   <!DOCTYPE html>
   <html>
       <head>
           <meta charset="utf-8" />
           <meta name="format-detection"
   content="telephone=no" />
           <meta name="msapplication-tap-highlight"
   content="no" />
           <meta name="viewport" content="user-scalable=no,
   initial-scale=1, maximum-scale=1, minimum-scale=1,
   width=device-width, height=device-height, target-
   densitydpi=device-dpi" />
           <link rel="stylesheet" type="text/css"
   href="jquery/jquery.mobile.min.css" />
           <script src="jquery/jquery.min.js"></script>
           <script src="jquery/jquery.mobile.min.js"></script>
           <title>Hello World</title>
       </head>
   ```

```
    <body>

        <script type="text/javascript"
src="cordova.js"></script>
        <script type="text/javascript">

        </script>
    </body>
</html>
```

5. Let's add the initial page for our application within the body of the HTML document. Here, we will create the page `div` element with the `id` attribute set to `contacts-home`:

```
<div data-role="page" id="contacts-home">

    <div data-role="header">
        <h1>My Contacts</h1>
    </div>

    <div data-role="content">


    </div>

</div>
```

6. Within the content `div` element, create a new unordered list block. Set the `id` attribute to `contactList`, `data-role` to `listview`, and the `data-inset` attribute to `true`:

```
<div data-role="content">
    <ul id="contactList" data-role="listview" data-inset="true">

    </ul>

</div>
```

7. With the HTML UI complete, let's now focus on creating the custom code to interact with the contacts. Create a new `script` tag block before the `body` close of the document, and include the event listener to check whether the device is ready, as well as the callback method the it will run—`onDeviceReady`:

```
<script type="text/javascript">
    document.addEventListener("deviceready", onDeviceReady,
false);

    function onDeviceReady() {
```

```
        getAllContacts();
    }
</script>
```

8. The application will execute the getAllContacts method, which will read from the device's contacts database. To achieve this, we'll set the optional multiple parameter of contactFindOptions to true so that we get multiple contacts.

> The multiple parameter is set to false by default, which will return only one contact.

9. We then set the required contactFields parameter to specify which fields should be returned in each Contact object.
10. Finally, we call the find() method, passing the fields, the options, and the success and error callback method names:

```
function getAllContacts() {
    var options = new ContactFindOptions();
    options.filter = "";
    options.multiple = true;
    var fields = ["name", "phoneNumbers",
        "birthday", "emails"
    ];
    navigator.contacts.find(fields,
        onAllSuccess, onError, options);
}
```

11. Following a successful response, the onAllSuccess method will return an array of Contact objects for us to work with. We will initially loop over the returned results and push each Contact object into a new array object, called arrContactDetails, which allows us to sort the results alphabetically. If no results are returned, we'll output a user-friendly message:

```
function onAllSuccess(contacts) {

    if (contacts.length) {

        var arrContactDetails = new Array();
        for (var i = 0; i < contacts.length; ++i) {
            if (contacts[i].name) {
                arrContactDetails.push(contacts[i]);
            }
```

```
        }

        arrContactDetails.sort(alphabeticalSort);

        // more code to go here

    } else {
        $('#contactList').append('<li><h3>Sorry,
            no contacts were found < /h3></li > ');
        }
        $('#contactList').listview("refresh");
    }
}
```

12. Include the `alphabeticalSort` function, which will sort each contact in ascending order using the formatted version of the name:

```
function alphabeticalSort(a, b) {
    if (a.name.formatted < b.name.formatted) {
        return -1;
    } else if (a.name.formatted > b.name.formatted) {
        return 1;
    } else {
        return 0;
    }
}
```

13. To create our contact list, the following code will go directly beneath the `arrContactDetails.sort(alphabeticalSort)` call in the code. This will loop over the sorted array and create the list items for each contact, setting the `Contact` object ID and the formatted name in each list item. It will also create the list divider to differentiate each group of contacts by the first letter of the name:

```
var alphaHeader = arrContactDetails[0].name.formatted[0];
for (var i = 0; i < arrContactDetails.length; ++i) {
    var contactObject = arrContactDetails[i];
    if (alphaHeader != contactObject.name.formatted[0]) {
        alphaHeader = contactObject.name.formatted[0];
        $('#contactList').append('<li data-role="list-divider">' +
alphaHeader + '</li>');
        $('#contactList').append(
            '<li class="contact_list_item" id="' +
            contactObject.id + '"><a href="#contact-info">'
+

            contactObject.name.formatted + ' (' +
```

```
                contactObject.id + ')</a></li>'
            );
        } else {
            if (i == 0) {
                $('#contactList').append(
                    '<li data-role="list-divider">' +
alphaHeader + '</li>');
            }
            $('#contactList').append(
                '<li class="contact_list_item" id="' +
contactObject.id + '"><a href="#contact-info">' +
                contactObject.name.formatted + ' (' +
                contactObject.id + ')</a></li>');
        }
}
```

14. Finally, we include the `onError` callback method, which will run if we encounter any issues obtaining data from the `find()` method:

```
function onError(error) {
    alert('An error has occurred: ' + error.code);
}
```

15. Upon running the application on a device, we will see the populated list somewhat like this:

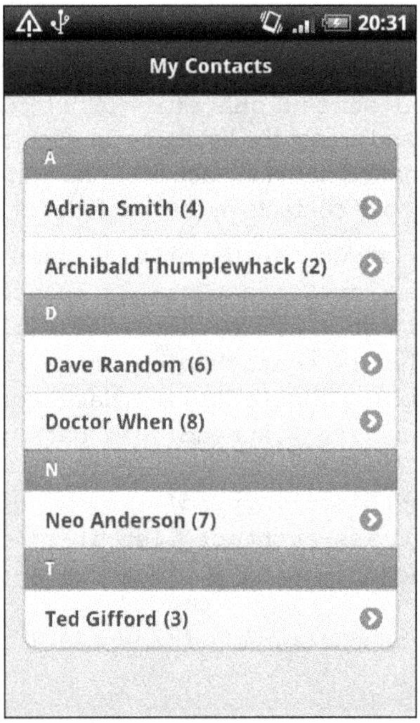

> Next to each name in the list, we can see the ID of the contact as used in the device's contact database. This value is also set in the `id` attribute of each list item.

How it works...

The `contacts.find()` method available from the contact plugin API is designed to query the device's contacts database to obtain and return an array of `Contact` objects. We set the contact fields in the required parameter of the function, which acts as a search qualifier for the transaction. Only the fields that we set in the `contactFields` parameter will be included as properties of the returned `Contact` objects. Using this parameter, we can choose exactly what details we want to obtain from the request for each contact.

Following a successful result from the `find()` method, an array of `Contact` objects is passed to the success callback method. Once we have received this information, we loop over the array to output the alphabetically sorted information in our unordered list, making use of the jQuery Mobile framework's `listview` item for clear display.

For a comprehensive look at the methods and properties available for use through the `Contact` object, refer to the official documentation at `http://plugins.cordova.io/#/package/org.apache.cordova.contacts`.

There's more...

In this example, we specifically set the values for the `contactFields` parameter to return in each `Contact` object. If this were left blank, we would receive only the `id` property of each contact. If we want to receive all available properties for each contact, we can set the value to a wildcard asterisk (`*`).

Displaying the contact information for a specific individual

Working with the contact database, developers can easily obtain a full array of all `Contact` objects saved on the device. We want to be able to obtain and view the saved contact information for specific individuals if we choose to drill down and filter a certain contact from the database.

Getting ready

For this recipe, we'll build on the code created in the previous recipe, *Listing all available contacts*. This will give us a head start to add more functionality to the application. Therefore, if you haven't yet completed the previous recipe, *Listing all available contacts*, it may help to complete it first.

How it works...

To manage the selected contact information, we will first make use of the `localStorage` API available and harness some more power from the jQuery Mobile framework:

1. When the user selects a contact from the list, we want to take them to a new page to show the details. Let's add a new page to `index.html` below the current one.
2. Set a new `div` element with the `data-role` attribute set to `page` and the `id` attribute set to `contact-info`. Within this, we will add our page header with the `id` attribute set to `contact_header`. We will also include a back button. It will take the user back to the original page by referencing the ID in the link.
3. We will keep the `H1` tag empty as we'll populate it with the contact's name:

```
<div data-role="page" id="contact-info">

    <div id="contact_header" data-role="header">
        <a href="#contacts-home" id="back" data-icon="back" 
data-direction="reverse">Back</a>
        <h1></h1>
    </div>

</div>
```

4. Below the page header, we create a content `div` element with the `id` attribute set to `contact_content`. It contains four form field items, which will display the given name, family name, phone number, and e-mail address for the chosen contact:

```
<div id="contact_content" data-role="content">

    <div data-role="fieldcontain">
        <label for="givenName">First Name:</label>
        <input type="text" name="givenName" id="givenName" 
disabled />
    </div>
    <div data-role="fieldcontain">
        <label for="familyName">Last Name:</label>
```

```
        <input type="text" name="familyName"
id="familyName" disabled />
    </div>
    <div data-role="fieldcontain">
        <label for="phone">Phone:</label>
        <input type="text" name="phone" id="phone" disabled
/>
    </div>
    <div data-role="fieldcontain">
        <label for="email">Email:</label>
        <input type="text" name="email" id="email" disabled
/>
    </div>

</div>
```

5. At the top of the custom JavaScript code, create a global variable to reference the `localStorage` API. We'll also include a global variable called `contactInfo`, which we will use to hold data later on:

```
<script type="text/javascript">
  var localStorage    =   window.localStorage;
  var contactInfo;

  document.addEventListener("deviceready", onDeviceReady,
false);
```

6. Let's now amend the `onAllSuccess` method, which writes the list of all contacts. Within the loop, we'll add a small portion of code that will add each item to `localStorage`. Here, we will store the entire contact object for each listing, and use the ID for each contact as the key that we can use to retrieve the information:

```
var alphaHeader = arrContactDetails[0].name.formatted[0];
for (var i = 0; i < arrContactDetails.length; ++i) {
    var contactObject = arrContactDetails[i];
    if (alphaHeader != contactObject.name.formatted[0]) {
        alphaHeader = contactObject.name.formatted[0];
        $('#contactList').append('<li data-role="list-divider">' +
alphaHeader + '</li>');
        $('#contactList').append(
            '<li class="contact_list_item" id="' +
            contactObject.id + '"><a href="#contact-info">'
+
            contactObject.name.formatted + ' (' +
            contactObject.id + ')</a></li>'
        );
```

```
        } else {
            if (i == 0) {
                $('#contactList').append(
                    '<li data-role="list-divider">' +
alphaHeader + '</li>');
            }
            $('#contactList').append(
                '<li class="contact_list_item" id="' +
contactObject.id + '"><a href="#contact-info">' +
                contactObject.name.formatted + ' (' +
                contactObject.id + ')</a></li>');
        }

        localStorage.setItem(
            contactObject.id,JSON.stringify(contactObject)
        );
    }
```

> The `localStorage` API saves data as a key/value pair and can only contain strings. As such, we convert the object to a string before saving it. For more information, check out the *Caching content using the local storage API* recipe in *Chapter 3, Filesystems, Storage, and Local Databases*.

7. Each of our generated list items references a particular contact. We have stored the specific ID for each contact as an attribute in each list item. Create an event handler that will obtain the value of the contact ID from the selected list item, and pass it to the `getContactByID` method:

```
$(document).on('click', '#contactList
li.contact_list_item', function() {

    var selectedID = $(this).attr('id');
    getContactByID(selectedID);

});
```

8. Let's now add the `getContactByID` function, which accepts the selected ID of the contact as a required parameter. This will obtain the selected contact information from `localStorage` and assign it to the `contactInfo` variable, which we set earlier. It will then take the user to a new page within the application:

```
function getContactByID(contactID) {
    contactInfo =
JSON.parse(localStorage.getItem(contactID));
```

```
        $.mobile.changePage($('#contact-info'));
    }
```

9. We now have the contact information stored, but we need to populate the form fields on the information page with the details. Let's add a new event handler to the code to detect a jQuery Mobile `pagechange` event, which will run a method called `onPageChange`:

```
$(document).bind("pagechange", onPageChange);
```

10. The `onPageChange` function will obtain the `id` attribute of the page that we have changed to. If it matches `contact-info`, we will first clear the values of all the form fields, and then set each one with the details from the `contactInfo` object. We will also set the `H1` tag in the header with the contact name:

```
function onPageChange(event, data) {
    var toPageId = data.toPage.attr("id");
    switch (toPageId) {
        case 'contact-info':

            clearValues();

            $('#contact_header h1')
                .html(contactInfo.name.formatted);
$('#givenName').val(contactInfo.name.givenName);
$('#familyName').val(contactInfo.name.familyName);

$('#phone').val(contactInfo.phoneNumbers[0].value);
            $('#email').val(contactInfo.emails[0].value);

            break;
    }
}
```

11. Finally, let's add the `clearValues` function, which will reset all form fields on the page where the input type is `text`:

```
function clearValues() {
    $('input[type=text]').each(function() {
        $('#' + this.id + '').val('');
    });
}
```

12. When we run the application on the device and select a contact, the resulting page will look like this:

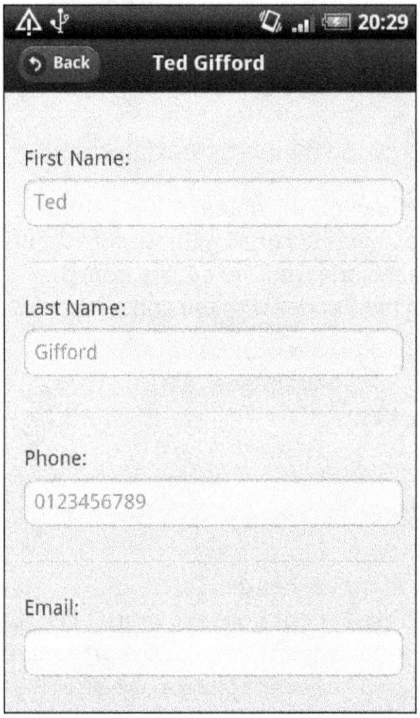

How it works...

In this recipe, we extended the previous code to fulfill the desired functionality. We amended the initial code when we looped over the returned array of `Contact` objects, and added some code to set each `Contact` object in the `localStorage` database available on the device, using its `ID` property as the key for the storage entry.

When a contact was selected from the list, we were able to use that list item's `ID` attribute to obtain the saved `Contact` object from `localStorage` before taking the user to the next page using the jQuery Mobile framework's `mobile.changePage()` function.

With the `Contact` object stored in an accessible variable, we were then able to read the properties set within it and output them to the user.

See also

- The *Caching content using the local storage API* recipe of *Chapter 3, Filesystems, Storage, and Local Databases*

Creating and saving a new contact

Having an address book or an application that can read contact information from the database is fantastic, but wouldn't it be even better if we could add contacts to the database? The good news is that the PhoneGap API not only provides a way to read the information, but also gives developers an incredibly powerful yet easy way to add information.

Getting ready

For this recipe, we'll build on the code created in the previous recipe, *Displaying the contact information of a specific individual*. This will give us a head start to add more functionality into the application. Therefore, if you haven't yet completed the previous recipe, *Displaying the contact information of a specific individual*, it would be a good idea to complete it first.

How to do it...

To store a new contact in the device database, we will create a form and save the new information in a `Contact` object:

1. Firstly, let's edit the default loading page for our application to include a button that will take the user to a new page to add a contact:

```
<div data-role="page" id="contacts-home">

    <div data-role="header">
        <h1>My Contacts</h1>
        <a href="#contact-add" id="back" data-icon="add">Add</a>
    </div>
```

2. Now let's create the new page. It will enable the user to input new contact information. Create a new jQuery Mobile page and set the `id` attribute to `contact-add`.
3. Within the header `div` element, we add a new link that will take the user back to the home page, bypassing the save functionality that we will shortly be adding:

```
<div data-role="page" id="contact-add">

    <div data-role="header">
        <a href="#contacts-home" id="back" data-icon="back"
data-direction="reverse">Back</a>
        <h1>Add Contact</h1>
```

```
        </div>

        <div data-role="content">

        </div>

    </div>
```

4. Within the content `div` element block, we will add a new `form` element with the `id` attribute set to `new_contact_form`. It contains a number of `form` field items. These will be used to enter the new information about the contact's given name, family name, phone number, and e-mail address.
5. The last form field block contains an input button with the `id` attribute set to `saveBtn`, which we'll reference via jQuery code to perform the save process.
6. Finally, we also include a hidden form item called `displayName`. We will populate this value after the form has been submitted, and will use it to store in the new `Contact` object:

```
<form id="new_contact_form">
    <div data-role="fieldcontain">
        <label for="givenName">First Name:</label>
        <input type="text" name="givenName" id="givenName"
/>
    </div>
    <div data-role="fieldcontain">
        <label for="familyName">Last Name:</label>
        <input type="text" name="familyName" id="familyName" />
    </div>
    <div data-role="fieldcontain">
        <label for="phone">Phone:</label>
        <input type="tel" name="phone" id="phone" />
    </div>
    <div data-role="fieldcontain">
        <label for="email">Email:</label>
        <input type="email" name="email" id="email" />
    </div>
    <div data-role="fieldcontain">
        <input type="button" name="saveBtn" id="saveBtn"
value="Save Contact" />
        <input type="hidden" name="displayName"
id="displayName" />
    </div>
</form>
```

7. With the layout and UI for the page complete, let's now focus on the JavaScript functionality for processing the new contact information. Amend the `onPageChange` function to add a new case within the `switch` statement to check for the page's `id` value `contact-add`. If it matches, everything within this `case` statement will be executable within that page context.

8. Firstly, we'll bind a `touchstart` event to the `saveBtn` button element, which will commence the saving process:

```
case 'contact-add':

    $('#saveBtn').bind('touchstart', function() {

    });

    break;
```

9. We can now populate the value of the `displayName` hidden form field by concatenating the values from the `givenName` and `familyName` form fields provided:

```
$('#saveBtn').bind('touchstart', function() {
    $('#displayName').val(
        $('#new_contact_form #givenName').val() + ' ' +
$('#new_contact_form #familyName').val()
    );

});
```

10. Before we can create the new `Contact` object, we must set the submitted information in the required format, and then we can send it as a parameter. The information is accepted in the form of a structural object containing key/value pairs. We can call each `form` field individually to create this, but in doing so, we would have a tightly coupled dependency on the specific `form` fields.

11. Here, we can make use of the jQuery library and create a serialized array of all the `form` fields within the form, which we have referenced by its `id` attribute.

> **Serialized array** means encoding a set of form elements as an array of names and values. For more information about serialized arrays, refer to `https://api.jquery.com/serializeArray/`.

12. We then loop over the array to create the key/value pairs (as expected) to return the structure of the information. Within the loop, we have set a `switch` statement to check for the name of the submitted form value. PhoneGap manages e-mails and phone numbers in a separate way from standard name contact fields. If they exist, we set them using a new `ContactField` object:

```
var arrContactInfo =
$('#new_contact_form').serializeArray();

var phoneNumbers = new Array();
var emails = new Array();

var contactInfo = '{';

for (var i = 0; i < arrContactInfo.length; i++) {
    switch (arrContactInfo[i].name) {
        case 'phone':
            if (arrContactInfo[i].value) {
                phoneNumbers[0] =
                    new ContactField('mobile',
                        arrContactInfo[i].value, true);
            }
            break;
        case 'email':
            if (arrContactInfo[i].value) {
                emails[0] =
                    new ContactField('work',
                        arrContactInfo[i].value, true);
            }
            break;
        default:
            contactInfo += '"' + arrContactInfo[i].name +
'" : "' + arrContactInfo[i].value + '"';
            if (i < arrContactInfo.length - 1) {
                contactInfo += ', '
            }
    }
}

contactInfo += '}';
```

13. Next, we create a new `Contact` object and pass the `contactInfo` variable that we just created before saving the contact in the device's database:

```
var newContact =
navigator.contacts.create(JSON.parse(contactInfo));

newContact.phoneNumbers = phoneNumbers;
```

```
newContact.emails = emails;

newContact.save(onSaveSuccess, onError);
```

14. While defining the `save()` method, we also include two callback methods to handle the successful save or any errors that may have arisen from the process. In the `onSaveSuccess` function, we will take the user back to the home page using the jQuery Mobile framework's built-in `changePage` method.
15. We will then refresh the `contactList` list element to show the new data stored within the device's database:

```
function onSaveSuccess(contact) {
    $.mobile.changePage($('#contacts-home'));
    $('#contactList').listview("refresh");
}
```

16. When we run the application on the device, the home page would look like this, with the new **Add** button:

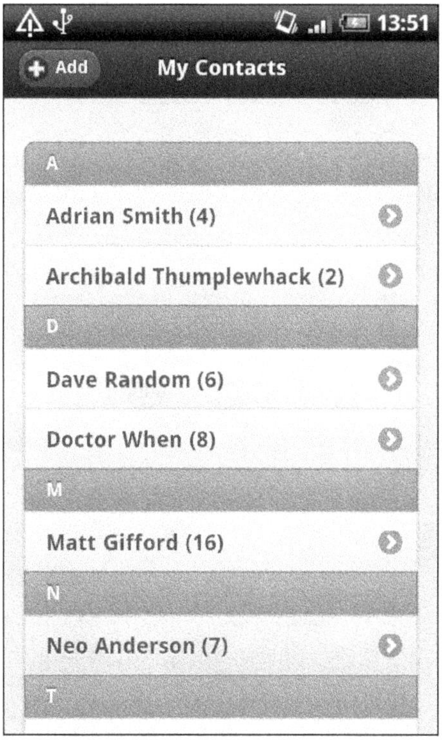

17. By choosing to add a new contact, the user will be presented with the new form page, as follows:

18. Finally, once the submission has been done, the user is taken back to the home page listing all contacts. It now shows the new contact, created and saved in the device database, like this:

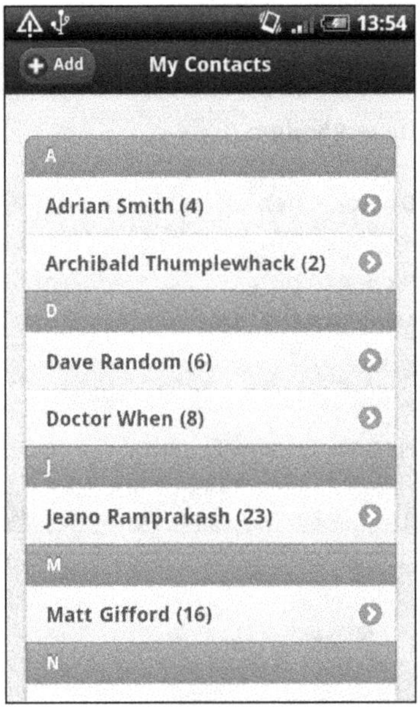

How it works...

When we create a new `Contact` object, we are simply creating a local variable populated with the information provided from the form submission. At this point, we can still add and amend properties within the `Contact` object, or simply fail to persist the object by not saving it.

To successfully save and store the contact within the device contact database, we then call the `save()` method, available from the `Contact` object.

The following methods are available for use from the `Contact` object:

- `save()`: This method will save a new contact in the device contacts database. If the `Contact` object has an `id` attribute that already matches that of a saved contact, it will update the saved contact with any revised information.
- `remove()`: Calling this method will remove the specified contact from the device contacts database.
- `clone()`: This method will create a deep copy of the provided `Contact` object. However, the `id` property will be set to `null`. This means that you can easily duplicate contact information and save it as a new `Contact` object.

There's more...

When we saved the e-mail address and phone number in the new `Contact` object, we used the `ContactField` object to do so instead of sending the values as part of the contact information object. The `ContactField` object is provided to support generic fields within a `Contact` object, such as e-mail addresses, phone numbers, and URLs.

The `Contact` object itself stores values such as these in an array, which can contain multiple `ContactField` objects—a contact can have more than one phone number assigned to it, for instance.

The `ContactField` object requires the following properties:

- `type`: This is a `DOMString` that identifies which type of field this is. For example, a phone number type value can include `home`, `work`, `mobile`, or any other value supported by the database on a particular device platform.
- `value`: This is a `DOMString` that holds the value of the field itself, for example, the phone number or e-mail address.
- `pref`: This is a `Boolean` value that, if set to `true`, will set the specific field as the preferred value for the `ContactField` type.

In our example application in this recipe, we stored only one phone number and one e-mail address per contact. We created a new array for each value in which to hold this information, but only set the first index in each array with a `ContactField` object.

Why not expand on this recipe to provide the user with a frontend UI that allows more form fields (in order to provide extra phone numbers), and then amend the code to create new `ContactField` objects for every new value submitted?

6
Hooking into Native Events

In this chapter, we will cover the following topics:

- Pausing your application
- Resuming your application
- Displaying the status of the device battery levels
- Displaying network connection status
- Creating a custom submenu

Introduction

When developing applications for mobile devices, we can create feature-rich applications that harness the functionality of the native processes and systems.

The devices themselves provide us with built-in controls and user interface elements in the form of native buttons, to which we can apply methods and functions.

We can also make use of the hidden events and manage how our applications will work when placed in the background of the device, or we can also alter states depending on network connectivity.

The recipes in this chapter will introduce you to some of the native events available through the PhoneGap API and how we can implement them in applications.

Pausing your application

Although we want our users to spend their time solely on our application, they will inevitably leave our application to open another one or do something else entirely. We need to be able to detect when a user has left our application but not closed it down entirely.

How to do it...

We can use the PhoneGap API to fire off a particular event when our application is in the background of the device:

1. Firstly, create a new PhoneGap project named `pausedemo` by running the following command:

   ```
   phonegap create pausedemo com.myapp.pausedemo pausedemo
   ```

2. Add the devices platform. You can choose to use Android, iOS, or both:

   ```
   cordova platform add ios
   cordova platform add android
   ```

3. Open `www/index.html` and let's clean up unnecessary elements. So we will have the following:

   ```
   <!DOCTYPE html>
   <html>
       <head>
           <meta charset="utf-8" />
           <meta name="format-detection"
   content="telephone=no" />
           <meta name="msapplication-tap-highlight"
   content="no" />
           <meta name="viewport" content="user-scalable=no,
   initial-scale=1, maximum-scale=1, minimum-scale=1,
   width=device-width, height=device-height, target-
   densitydpi=device-dpi" />
           <title>Hello World</title>
       </head>
       <body>

           <script type="text/javascript"
   src="cordova.js"></script>
           <script type="text/javascript">

           </script>
       </body>
   </html>
   ```

4. Before the closing the `body` tag, add the event listener inside the `script` tag block to check when the device is ready and the PhoneGap code is ready to run:

```
<script type="text/javascript">
    document.addEventListener("deviceready",
        onDeviceReady, false);
</script>
```

5. Create the `onDeviceReady` function, which will run when the event listener is fired. Inside this, we'll create a new event listener that will check for a `pause` event, and once received, this will fire the `onPause` method:

```
function onDeviceReady() {
    document.addEventListener("pause", onPause, false);
}
```

6. Let's create the `onPause` method. In this example application, we'll ask the device by playing an audio beep to notify the user that the application has moved to the background. The numeric parameter specifies how many times we want the audio notification to be played—in this case, just once:

```
function onPause() {
    navigator.notification.beep(1);
}
```

> Developing for iOS? There is no native beep API for iOS. The PhoneGap API will play an audio file using the media API, but the developer must provide the file named `beep.wav` in the `/www` directory of the application project files. This file should be under 30 seconds in length. iOS will also ignore the beep count argument and will play the audio once. In Windows 7 mobile, the WP7 Cordova library contains a generic beep audio file that will be used if the application is developed for this OS.

7. When running the application on the device, if you press the home button or navigate to another application, the device will play the notification audio.

How it works...

To correctly determine the flow of our lifecycle events, we have first set up the `deviceready` event listener to ensure that the native code was properly loaded. At this point, we were then able to set the new event listener for the `pause` event.

As soon as the user navigates away from our application, the native code would set the application as a background processes on the device and fire the `pause` event, at which point our listener would run the `onPause` method.

To find out more about the `pause` event, refer to the official documentation available here:

`http://docs.phonegap.com/en/3.6.0/cordova_events_events.md.html#pause`

There's more...

In this recipe, we applied the `pause` event in an incredibly simple manner. There is a possibility that your application will want to do something specific other than sending an audio notification when the user pauses your application.

For example, you may want to save and persist any data currently in the view or in memory, such as any draft work (if dealing with form inputs) or saving responses from a remote API call.

We'll build an example that will persist data in the next recipe, as we'll be able to quantify its success when we resume the use of the application and bring it back to the foreground.

Resuming your application

Multi-tasking capabilities that are now available on mobile devices specify that the user has the ability to switch from one application to another at any time. We need to handle this possibility and ensure that we can save and restore any process and data when the user returns to our application.

How to do it...

We can use the PhoneGap API to detect when our application is brought back to the foreground of the device:

1. Firstly, create a new PhoneGap project named `resumedemo` by running the following command:

   ```
   phonegap create resumedemo com.myapp.resumedemo resumedemo
   ```

2. Add a devices platform. You can choose to use Android, iOS, or both:

   ```
   cordova platform add ios
   cordova platform add android
   ```

3. Open `www/index.html` and let's clean up unnecessary elements. We'll also be manipulating the DOM elements, so include a reference to the `xui.js` file within the `head` tag. We will also be setting the `deviceready` listener once the DOM has fully loaded, so let's apply an `onload` attribute to the `body` tag:

   ```
   <!DOCTYPE html>
   <html>
       <head>
   ```

```
        <meta charset="utf-8" />
        <meta name="format-detection"
content="telephone=no" />
        <meta name="msapplication-tap-highlight"
content="no" />
        <meta name="viewport" content="user-scalable=no,
initial-scale=1, maximum-scale=1, minimum-scale=1,
width=device-width, height=device-height, target-
densitydpi=device-dpi" />
        <script type="text/javascript"
src="js/xui.js"></script>
        <title>Hello World</title>
    </head>
    <body onload="onLoad()">


        <script type="text/javascript"
src="cordova.js"></script>
    </body>
</html>
```

4. Create a new `script` tag block before the closing `body` tag and add the `deviceready` event listener within the `onLoad` method. We'll also set two global variables, `savedTime` and `localStorage`; the latter of which will reference the `localStorage` API on the device:

```
<script type="text/javascript">
    var savedTime;
    var localStorage    =   window.localStorage;

    function onLoad() {
            document.addEventListener("deviceready",
            onDeviceReady, false);
    }
</script>
```

5. Create the `onDeviceReady` function, within which we'll set the two event listeners to check for the `pause` and `resume` events:

```
function onDeviceReady() {
    document.addEventListener("pause", onPause, false);
    document.addEventListener("resume", onResume, false);
}
```

6. We can now add the first of the new callback functions for the added listeners. `onPause` will run when a `pause` event has been detected. In this method, we'll create a new date variable holding the current time and store it in the global `savedTime` variable we created earlier.

7. If the user has entered something in the text input field, we'll take this value also and send it to the `localStorage` API before clearing out the input field:

```
function onPause() {
    savedTime = new Date();
    var strInput = x$('#userInput').attr('value');
    if(strInput) {
        localStorage.setItem('saved_input', strInput);
        x$('#userInput').attr('value', '');
    }
}
```

8. Define the `onResume` method, which will run when a `resume` event has been detected. In this function, we'll save a new date variable and we'll use it in conjunction with the `savedTime` variable created in the `onPause` method to generate the time difference between the two dates. We'll then create a string message to display the time details to the user.

9. We'll then check `localStorage` for the existence of an item stored using the `saved_input` key. If this exists, we'll extend the message string and append the saved user input value before sending the message to the DOM to display:

```
function onResume() {
    var currentTime = new Date();
    var dateDiff = currentTime.getTime() -
savedTime.getTime();
    var objDiff = new Object();
    objDiff.days = Math.floor(dateDiff / 1000 / 60 / 60 /
24);
    dateDiff -= objDiff.days * 1000 * 60 * 60 * 24;
    objDiff.hours = Math.floor(dateDiff / 1000 / 60 / 60);
    dateDiff -= objDiff.hours * 1000 * 60 * 60;
    objDiff.minutes = Math.floor(dateDiff / 1000 / 60);
    dateDiff -= objDiff.minutes * 1000 * 60;
    objDiff.seconds = Math.floor(dateDiff / 1000);

    var strMessage = '<h2>You are back!</h2>'
    strMessage += '<p>You left me in the background for '
    strMessage += '<b>' + objDiff.days + '</b> days, '
    strMessage += '<b>' + objDiff.hours + '</b> hours, '
    strMessage += '<b>' + objDiff.minutes + '</b> minutes,
'
    strMessage += '<b>' + objDiff.seconds + '</b>
seconds.</p>';
```

```
    if (localStorage.getItem('saved_input')) {
        strMessage = strMessage + '<p>You had typed the
following before you left:<br /><br />'
        strMessage += '"<b>' +
localStorage.getItem('saved_input') + '</b>"</p>';
    }

    x$('#message').html(strMessage);

}
```

10. Finally, let's add the DOM elements to the application. Create a new `div` element with the `id` attribute set to `message`, and an `input` text element with the `id` set to `userInput`:

```
<body onload="onLoad()">
    <div id="message"></div>

    <input type="text" id="userInput" />
```

11. If we run the application on the device, the initial output would provide the user with an input box to enter text, should they wish to:

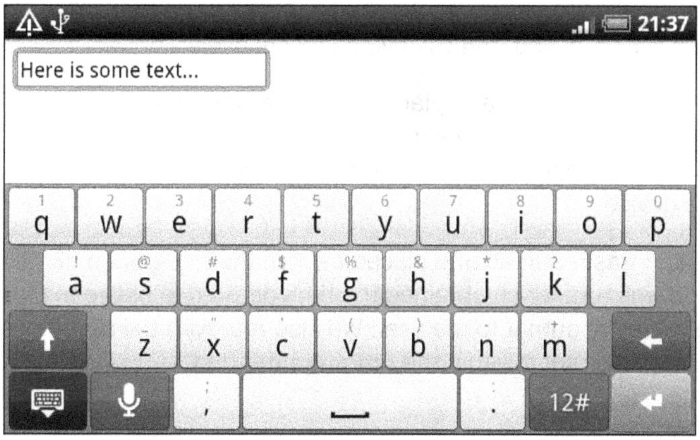

12. If we were to pause the application and then resume it after some time, the display would then update to look something like this:

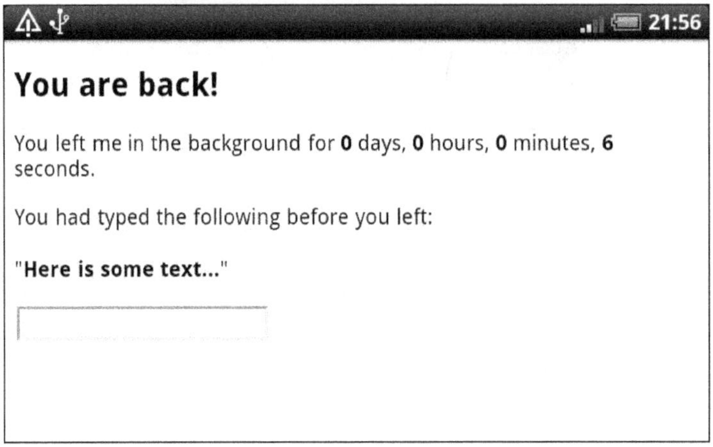

How it works...

We have set up the `deviceready` event listener after the DOM was fully loaded, which would then run the `onDeviceReady` function. Within this method, we then added two new event listeners to catch the `pause` and `resume` events respectively.

When the application was paused and placed in the background processes of the device, we saved the current date and time in a global variable. We also checked for the existence of any user-supplied input, and if it was present, we saved it using the `localStorage` capabilities of the device.

When the application was resumed and placed back in the foreground of the device, the `onResume` method was run, which obtained the time difference between the saved and current datetime values to output to the user. We also retrieved the saved user input from `localStorage`, if we had set it within the `onPause` method.

> To find out more about the `resume` event, refer to the official documentation available here: `http://docs.phonegap.com/en/3.6.0/cordova_events_events.md.html#resume`.

Displaying the status of the device battery levels

Progression in capabilities and processing power means we can do much more with our mobile devices including multitasking and background processes. This often means we end up using more battery power to fuel our applications.

How to do it...

In this recipe, we will build an application to display the connection details and the current power capacity of the device battery:

1. Firstly, create a new PhoneGap project named `batterystatus` by running the following:

   ```
   phonegap create batterystatus com.myapp.batterystatus
   batterystatus
   ```

2. Add a devices platform. You can choose to use Android, iOS, or both using the following command:

   ```
   cordova platform add ios
   
   cordova platform add android
   ```

3. We will use a `battery-status` plugin:

   ```
   cordova plugin add org.apache.cordova.battery-status
   ```

4. Open `www/index.html` and let's clean up unnecessary elements. We'll also manipulate the DOM elements, so include a reference to the `xui.js` file within the `head` tag.
5. We will be calling the `onDeviceReady` method to instantiate the PhoneGap functionality through an `onLoad()` function attached to the `body` tag:

   ```
   <!DOCTYPE html>
   <html>
       <head>
           <meta charset="utf-8" />
           <meta name="format-detection"
   content="telephone=no" />
           <meta name="msapplication-tap-highlight"
   content="no" />
           <meta name="viewport" content="user-scalable=no,
   initial-scale=1, maximum-scale=1, minimum-scale=1,
   width=device-width, height=device-height, target-
   densitydpi=device-dpi" />
   ```

```
        <script type="text/javascript"
src="js/xui.js"></script>
        <title>Hello World</title>
    </head>
    <body onload="onLoad()">

        <script type="text/javascript"
src="cordova.js"></script>
        <script type="text/javascript">

        </script>
    </body>
</html>
```

6. Let's add the UI elements for the application. This will include a `div` element with the `id` attribute set to `statusMessage`, which will hold our returned information. We'll also build up some nested elements to create a visual representation of the device battery.
7. We will reference the `id` attributes of each element using the `XUI` library, so we need to make sure the three attribute values are set to `batteryIndicator`, `batteryLevel`, and `shade` respectively:

```
<h3>Battery Status</h3>

<div id="statusMessage"></div>

<div id="batteryIndicator">
    <div id="batteryLevel">
        <div id="shade" />
    </div>
</div>
```

8. With the DOM elements inserted, let's move on to the JavaScript code. Create a `script` tag block before the closing `head` tag and add an `onLoad` method that is run from the `body` tag, which will add the event listener to check whether the native PhoneGap code has been loaded and is ready for use:

```
<script type="text/javascript">
    function onLoad() {
        document.addEventListener("deviceready",
          onDeviceReady, false);
      }
</script>
```

9. Add the `onDeviceReady` method, to which we will add three new event listeners that will respond to the changes in the device's battery status. Each listener has a corresponding callback method, which we will define in the next few steps:

```
function onDeviceReady() {
    window.addEventListener("batterystatus",
    onBatteryStatus, false);
    window.addEventListener("batterylow", onBatteryLow,
false);
    window.addEventListener("batterycritical",
    onBatteryCritical, false);
}
```

10. The first callback method is `onBatteryStatus`, which accepts the information object that contains the properties of the device's battery. We will pass this information to a new function, `setBatteryInfo`:

```
function onBatteryStatus(battery_info) {
    setBatteryInfo(battery_info);
}
```

11. Let's write the `setBatteryInfo` method called from the status change function. We can use the `level` property returned from the `battery_info` object to set the width of the `batteryLevel` element. We'll then create a message with the current capacity level and stating whether or not the device is plugged in. This message will be sent to the `statusMessage` element.

12. If the battery level is below 21 percent, we'll change the background color of the battery to red, otherwise we'll set it to a healthy green:

```
function setBatteryInfo(battery_info) {
    x$('#batteryLevel').setStyle('width',
        battery_info.level + '%');
    var statusMessage = '<p>Percent: <span id="level">' +
        battery_info.level + '%</span></p>';
    statusMessage = statusMessage + '<p>A/C: ' +
        chargingStatus(battery_info.isPlugged) + '</p>';
    x$('#statusMessage').html(statusMessage);
    if (battery_info.level <= 20) {
        x$('#level').addClass('warning');
        x$('#batteryLevel').setStyle('backgroundColor',
            '#E74A4A');
    } else {
        x$('#batteryLevel').setStyle('backgroundColor',
            '#01A206');
    }
}
```

13. The return value of the `isPlugged` value from the `battery_info` object is a `Boolean` value, so we'll send it to a new function to return a string representation of the connection:

```
function chargingStatus(isPlugged) {
    if(isPlugged) { return 'Connected'; }
    return 'Disconnected';
}
```

14. The `batterylow` event handler will run a method called `onBatteryLow`. Inside this, we'll include a notification alert to inform the following to the user:

```
function onBatteryLow(battery_info) {
    navigator.notification.alert(
        'Time to charge it up!',
        function() {}, //alert dismissed
        'Low Battery',
        'OK'
    );
}
```

15. If the battery reaches critical levels, the `onBatteryCritical` callback method will run. Again, let's use this event to alert the user of the urgent need to charge the device:

```
function onBatteryCritical(battery_info) {
    navigator.notification.alert(
        'Seriously, plug your charger in!',
        function() {}, //alert dismissed
        'Critical Battery',
        'OK'
    );
}
```

16. Both our notification alerts will execute an empty function specified as the `alertCallback` property in the previous snippets, which will run when the alert is dismissed. For this example, we don't need to perform any extra functionality at this point, hence the empty function.

17. Finally, include some CSS definitions to add a visual presence to our battery elements:

```
<style>
#batteryIndicator {
    margin: 0 auto;
    width: 250px;
    height: 100px;
    border: 1px solid #ccc;
```

```
        background: #fff;
        border-radius: 10px;
        overflow: hidden;
    }
    #batteryLevel {
        height: 100%;
    }
    #shade {
        width: 100%;
        height: 15px;
        background: -webkit-gradient(linear, 0% 0%, 0% 100%,
    from(#e5e5e5), to(#fff));
        opacity: 0.2;
        position: relative;
        top: 15px;
    }
    .warning {
        color: #ff0000;
        font-weight: bold;
    }
    </style>
```

18. When running the application on the device, assuming that the battery levels are within the healthy boundaries, the output will look similar to this:

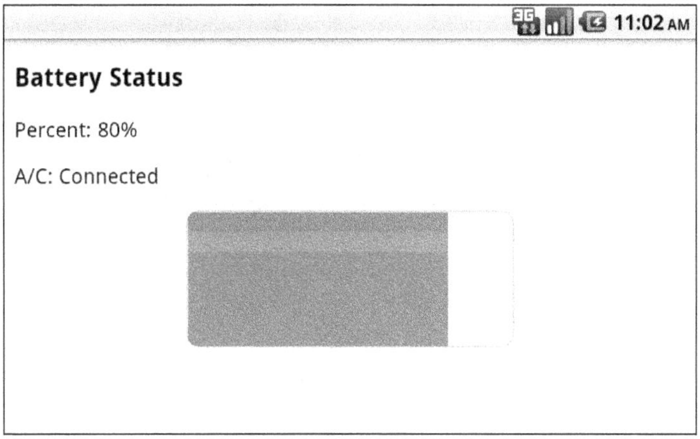

19. As soon as the device battery capacity goes below 21 percent, the UI will change to something similar to the following:

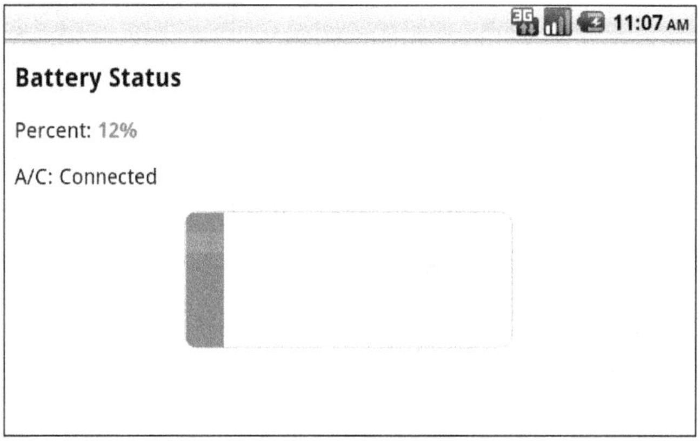

20. In this image, the device battery levels have hit the critical value of **5%**. As a result, we notify the user with an alert message:

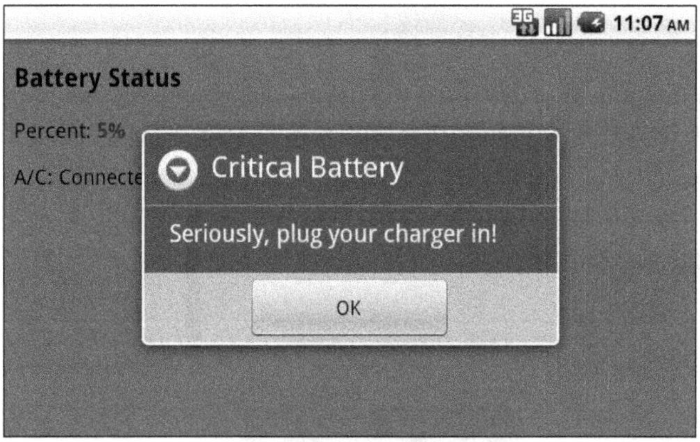

How it works...

Using the `onDeviceReady` method, we have set up three new event listeners to check for the status of the device battery levels.

All three battery status handlers (`batterystatus`, `batterylow`, and `batterycritical`) return the same object with the following properties:

- `level`: `Number` that defines the percentage of the battery, between 0 and 100
- `isPlugged`: `Boolean` that returns a true or false value to represent whether the device is connected to a charger or not

We were able to use the returned properties to update a visual representation of the device battery on screen, as well as use the `level` property to determine whether we are below the low threshold barrier.

Although at first glance, the three events seem to do the same thing, there are important differences. The `batterystatus` event will detect changes in the battery capacity and will fire its callback method with every percentage change. This allows us to keep a constant check on the status of the device battery levels. It will also fire if the device is connected or disconnected from the charger. From the `isPlugged` property, we can easily determine whether the device is using the mains' power or not.

The `batterylow` event will fire only when the battery has reached a specific percentage level deemed as low by the device. The same applies to the `batterycritical` event, which will only fire once the battery level has reached a particular percentage.

> The threshold levels for the `batterylow` and `batterycritical` events are specific to each device, so this is something to be aware of if you are hardcoding values within the application. As a reference, Android devices typically set the low threshold to 20 percent and the critical threshold to 5 percent.

To find out more about the `batterycritical`, `batterylow`, and `batterystatus` events, refer to the official documentation available here:

`https://github.com/apache/cordova-plugin-battery-status.`

There's more...

In our sample application included in this recipe, we processed a simple alert notification when the low and critical thresholds were reached.

Depending on your mobile application and what its processes are, the chances are that you will want to do something specific at these points. For example, you may want to save any user input values to the local memory or shutdown/pause certain aspects of the application's functionality once these thresholds have been detected. If the user is unable to charge their device, you do not want them to lose data while using your application.

Displaying network connection status

Your application may require the user to be connected to a network. This may be for partial updates, remote data transfer, or streaming. Using the PhoneGap API and network information plugin, we can easily detect the status or existence of any network connectivity.

How to do it...

In this recipe, we will build an application to constantly check the network connection status of our device:

1. Firstly, create a new PhoneGap project named `networkinfo` by running the following command:

   ```
   phonegap create networkinfo com.myapp.networkinfo networkinfo
   ```

2. Add the devices platform. You can choose to use Android, iOS, or both:

   ```
   cordova platform add ios
   cordova platform add android
   ```

3. We will use the network info plugin as follows:

   ```
   cordova plugin add org.apache.cordova.network-information
   ```

4. Open `www/index.html` and let's clean up unnecessary elements. We'll also be manipulating the DOM elements, so include a reference to the `xui.js` file within the `head` tag.
5. We will also set the `deviceready` event listener after the DOM has fully loaded, so we'll also add the `onLoad()` function call to the `body` tag:

   ```
   <!DOCTYPE html>
   <html>
       <head>
           <meta charset="utf-8" />
           <meta name="format-detection"
   content="telephone=no" />
           <meta name="msapplication-tap-highlight"
   content="no" />
           <!-- WARNING: for iOS 7, remove the width=device-width and
   height=device-height attributes. See
   https://issues.apache.org/jira/browse/CB-4323 -->
           <meta name="viewport" content="user-scalable=no,
   initial-scale=1, maximum-scale=1, minimum-scale=1,
   width=device-width, height=device-height, target-
   densitydpi=device-dpi" />
   ```

```
        <script type="text/javascript"
src="js/xui.js"></script>
        <title>Hello World</title>
    </head>
    <body onload="onload()">

        <script type="text/javascript"
src="cordova.js"></script>
        <script type="text/javascript">

        </script>
    </body>
</html>
```

6. Let's add the UI elements to the body of our application. Create a `div` block to act as a container for our `statusMessage` and `count` elements, both of which we will reference directly using the XUI library. We will also insert content into the `speedMessage` element, so ensure the `id` attribute of these three elements match those shown:

```
<h3>Network Status</h3>

<div id="holder">

    <div id="statusMessage"></div>

    <div id="count"></div>

</div>

<div id="speedMessage"></div>
```

7. Create a new `script` tag block before the closing `body` tag and define two global variables, which we will use within the custom code. We can also now define the `onLoad` method, which will set the `deviceready` event listener:

```
<script type="text/javascript">
    var intCheck = 0;
    var currentType;

    function onLoad() {
        document.addEventListener("deviceready",
            onDeviceReady, false);
    }

</script>
```

8. Let's now add the `onDeviceready` method called from the `deviceready` event listener. Within this function, we will add two new event listeners to check when the device is connected or disconnected from a network. Both of these listeners will run the same callback method, `checkConnection`.

9. We will then set up an interval timer to run the same `checkConnection` method every second to provide us with constant updates for the connection:

```
function onDeviceReady() {
    document.addEventListener("online", checkConnection,
        false);
    document.addEventListener("offline", checkConnection,
        false);
    var connCheck = setInterval(function() {
        checkConnection();
    }, 1000);
}
```

10. The `checkConnection` function sets up the `objConnection` variable to hold a representation of the device's connection. This object returns a value in the `type` property from which we are able to determine the current connection type. We'll pass this value to another function called `getConnectionType`, which we'll use to return a user-friendly string representation of the connection type.

11. As this method runs every second, we want to be able to determine whether the current connection type differs from the previous connection. We can do this by storing the connection type value in the `currentType` global variable and check whether this matches the current value.

12. Depending on the returned value of the connection type, we can alternatively choose to inform the user that to get the most out of our application they should have a better connection.

13. We will also increment an integer value stored in the `intCheck` global variable, which we will use to count the number of seconds the current connection has been active:

```
function checkConnection() {
    var objConnection = navigator.network.connection;
    var connectionInfo = getConnectionType(objConnection.type);
    var statusMessage = '<p>' + connectionInfo.message +
'</p>';

    if(currentType != objConnection.type) {
        intCheck = 0;
        currentType = objConnection.type;
```

```
        if(connectionInfo.value <= 3) {
            x$('#speedMessage').html('<p>This application
works better over a faster connection.</p>');
        } else {
            x$('#speedMessage').html('');
        }
    }
    intCheck = intCheck + 1;

    x$('#statusMessage').html(statusMessage);
    x$('#count').html('<p>Checked ' + intCheck + ' seconds
ago</p>');
}
```

14. The `getConnectionType` method mentioned previously will return a `message` and a `value` property depending on the `type` value sent as the parameter. The `value` properties have been assigned manually to allow us to control what level of connection we deem best for our application and for the experience of our users:

```
function getConnectionType(type) {
    var connTypes = {};
    connTypes[Connection.NONE] = {
        message: 'No network connection',
        value: 0
    };
    connTypes[Connection.UNKNOWN] = {
        message: 'Unknown connection',
        value: 1
    };
    connTypes[Connection.ETHERNET] = {
        message: 'Ethernet connection',
        value: 2
    };
    connTypes[Connection.CELL_2G] = {
        message: 'Cell 2G connection',
        value: 3
    };
    connTypes[Connection.CELL_3G] = {
        message: 'Cell 3G connection',
        value: 4
    };
    connTypes[Connection.CELL_4G] = {
        message: 'Cell 4G connection',
        value: 5
    };
    connTypes[Connection.WIFI] = {
```

```
            message: 'WiFi connection',
            value: 6
        };
        return connTypes[type];
    }
```

15. Finally, let's add some CSS definitions to the bottom of our application to add some style to the UI:

```
<style>
div#holder {
    width: 250px;
    min-height: 60px;
    margin: 0 auto;
    position: relative;
    border: 1px solid #ff0080;
    border-radius: 10px;
    background: #ff0080;
}
div#holder p {
    margin: 20px auto;
    text-align: center;
    color: #ffffff;
    font-weight: bold;
}
div#speedMessage {
    width: 250px;
    margin: 0 auto;
    position: relative;
}
</style>
```

16. On running the application on our device, the output will look something like this:

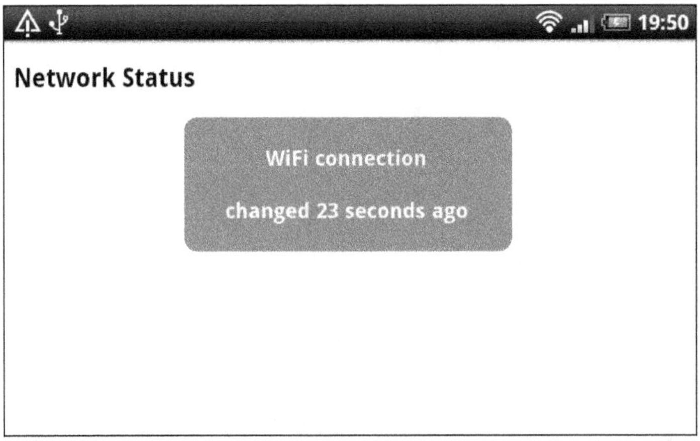

17. If our user changes their connection method or disconnects completely, the interval timer will detect the change and update the interface accordingly, and the timer will restart:

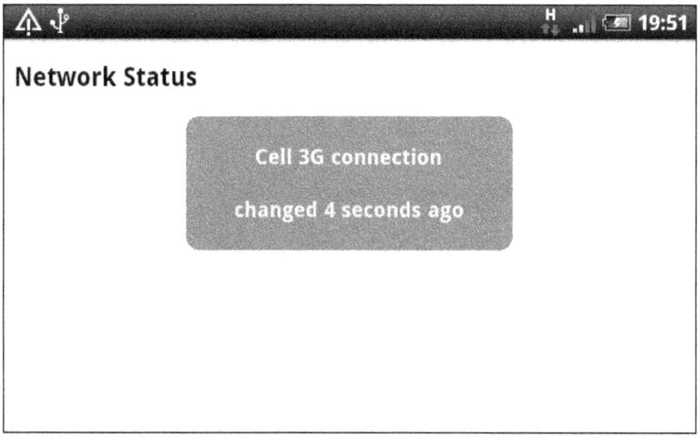

How it works...

We have set up the `onDeviceReady` method to create two new event listeners to check for the `online` and `offline` events respectively.

> The `online` event will fire when the device's network connection is started, and the `offline` event will fire when the network connection is turned off or lost.

These events will only fire once, and so in this recipe we added the `setInterval` timer function to constantly call the `checkConnection` method to allow us to obtain the changes made to the network. The addition of this functionality helps greatly and means that we can tell when a user switches from a 3G to a Wi-Fi connection, for example. If this happens, they would not go offline, but simply change the connection type.

> To find out more about the `online` and `offline` events, refer to the official documentation available here:
>
> `https://github.com/apache/cordova-plugin-battery-status`.

There's more...

Your application may involve streaming data, remote connections, or another process that requires a certain level of connectivity to a network. By constantly checking the status and the type of connection, we can determine whether it falls below an optimal level or it is the recommended type for your application. At this point, you could inform the user or restrict access to certain remote calls or data streams to avoid latency in your application's response and possible extra financial costs incurred to the user from their mobile provider.

Creating a custom submenu

Your application may include an option for users to update or change settings, or perhaps the ability to truly exit the application gracefully by closing down all services and storing state or data.

How to do it...

In this recipe, we will create a simple application that interacts with the device's native menu button to create a sliding submenu:

1. Firstly, create a new PhoneGap project named `submenu` by running the following command:

   ```
   phonegap create submenu com.myapp.submenu submenu
   ```

2. Add a devices platform. You can choose to use Android as follows:

   ```
   cordova platform add android
   ```

3. Open `www/index.html` and let's clean up unnecessary elements. We'll also manipulate the DOM elements, so include a reference to the `xui.js` file within the `head` tag. We will include an `onLoad` method call within the `body` tag, which will set the `deviceready` event listener once the DOM is fully loaded:

   ```
   <!DOCTYPE html>
   <html>
       <head>
           <meta charset="utf-8" />
           <meta name="format-detection"
   content="telephone=no" />
           <meta name="msapplication-tap-highlight"
   content="no" />
           <meta name="viewport" content="user-scalable=no,
   initial-scale=1, maximum-scale=1, minimum-scale=1,
   width=device-width, height=device-height, target-
   densitydpi=device-dpi" />
   ```

```
        <title>Hello World</title>
    </head>
    <body onload="onLoad()">

        <script type="text/javascript"
src="cordova.js"></script>
        <script type="text/javascript"
src="js/xui.js"></script>
        <script type="text/javascript">

        </script>
    </body>
</html>
```

4. Let's now add the UI to the `body` of the document. Create a new `button` element with the `id` attribute set to `menuToggle` and an unordered list within a `div` element. In this recipe, each `anchor` tag has a specific `id` attribute, which we'll use shortly to assign touch handlers to each link:

```
<h2>Menu Display</h2>

<button id="menuToggle">Toggle Menu</button>

<div id="subMenu">
    <a id="closeMenu"><span>Close Menu</span></a>
    <a id="hello"><span>Hello</span></a>
    <a id="exit"><span>Exit Application</span></a>
</div>
```

5. Add a new `script` tag block before the closing `head` tag in the document and include the `onLoad` function that will add the `deviceready` event listener:

```
<script type="text/javascript">
    function onLoad() {
    document.addEventListener("deviceready",onDeviceReady,
    false);
    }

</script>
```

6. Create the `onDeviceReady` method, within which we will set a new event listener to check for the `menubutton` event. This will run the `onMenuPress` function once it is detected.

7. We'll also include the `setMenuHandlers` method, which will apply the touch handlers to the menu items:

```
function onDeviceReady() {
    document.addEventListener("menubutton", onMenuPress,
false);
    setMenuHandlers();
}
```

8. The `onMenuPress` function, that is the callback from the event listener, will handle the transition of our menu element and links. We will use the XUI library to determine the current value of the `subMenu` element position and react accordingly to either open or close the menu:

```
function onMenuPress() {
    var menuPosition = x$('#subMenu').getStyle('bottom');
    if (menuPosition == '-100px') {
        x$('#subMenu').tween({
            bottom: '0px'
        });
    } else {
        x$('#subMenu').tween({
            bottom: '-100px'
        });
    }
}
```

9. The `setMenuHandlers` method will apply the touch handlers to our individual menu items. We can reference each element by the `id` attribute and set the listener with the specific action we want it to run. To exit the application, we can call the `exitApp` method to gracefully close our application and not leave it running in the background of the device.

10. The `menuToggle` button element and the `closeMenu` link item both provide the user with the ability to close the menu by calling the previously created `onMenuPress` method:

```
function setMenuHandlers() {
    x$('#exit').on('touchstart', function() {
        navigator.app.exitApp();
    });
    x$('#hello').on('touchend', function() {
        alert('Hello!');
    });
    x$('#closeMenu').on('touchend', function() {
        onMenuPress();
    });
```

```
    x$('#menuToggle').on('touchend', function() {
        onMenuPress();
    });
}
```

11. Finally, we will include some CSS definitions to set the menu into position and apply color styles and required attributes:

```
<style>
#menuToggle {
    width: 100%;
    height: 40px;
    position: relative;
    margin: 0 auto;
    color: #ffffff;
    background: #ff0080;
    font-size: 20px;
}
#subMenu {
    position: fixed;
    bottom: -100px;
    left: 0px;
    border-top: 1px solid #555;
    height: 100px;
    width: 100%;
    background: #e5e5e5;
}
#subMenu span {
    position: relative;
    margin: 0 auto;
    top: 40%;
}
#subMenu a {
    width: 29%;
    height: 100px;
    display: block;
    float: left;
    margin: 0 2% 0 2%;
    border-left: 1px solid #ccc;
    border-right: 1px solid #ccc;
    text-align: center;
}
</style>
```

12. When the application is run on the device, the initial view will look like this:

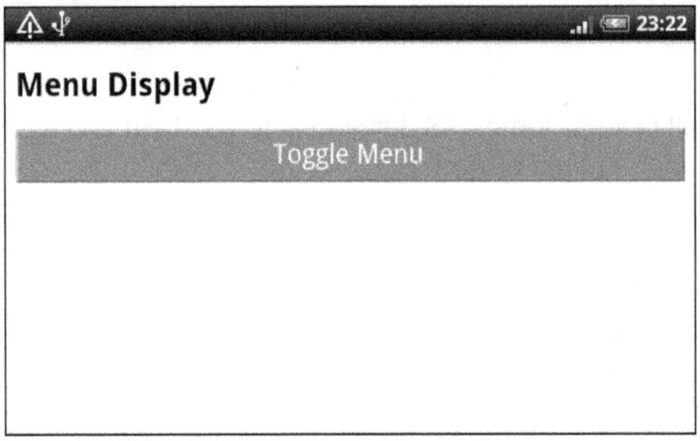

13. With the menu open, the user will be presented with our links as follows:

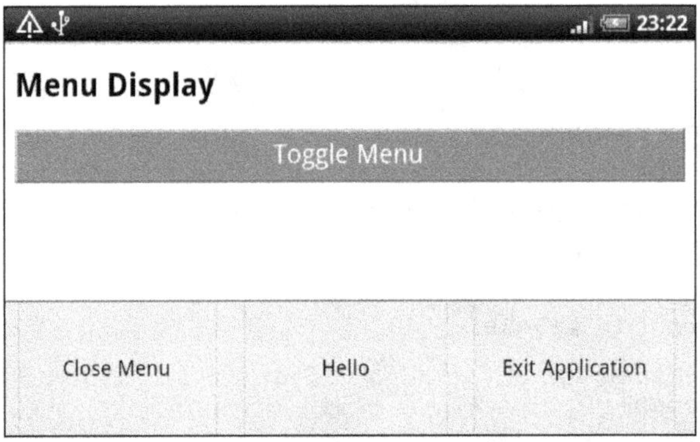

How it works...

The `onDeviceReady` method set up the new event listener to listen for the `menubutton` event. At this point, the `onMenuPress` function is run, which either opens or closes the menu depending on the current position of the `subMenu` element.

This is an ideal way to incorporate menu options and the hidden gems of functionality within your application without overcrowding your user interface.

> To create the transition of the menu, we have used the Tween capabilities provided by the XUI JavaScript library. We'll cover the library in more detail in the next chapter.
>
> To find out more about the `menubutton` event, refer to the official documentation available here: `http://docs.phonegap.com/en/3.6.0/cordova_events_events.md.html#menubutton`.

There's more...

The `menubutton` event provided by the PhoneGap API is not cross-device or cross-platform capable. At present, the supported device platforms are Android and BlackBerry WebWorks (OS 5.0 or higher).

There are other ways to include custom menus in your applications, thanks to one of the many PhoneGap plugins created by the community developers and users.

The Native Menu plugin (`https://github.com/mwbrooks/cordova-plugin-menu`) allows you to add native toolbars, tabbars and menus to your application and is supported on Android, BlackBerry WebWorks, and iOS platforms.

The community and open source nature of the PhoneGap API and the Cordova product means that developers can freely extend and enhance the functionality of their applications and dig a little deeper into native processes offered by the devices by creating custom plugins.

See also

- *Chapter 11, Extending PhoneGap with Plugins*

7
Working with XUI

In this chapter, we will cover the following recipes:

- Learning the basics of the library
- DOM manipulation
- Working with touch and gesture events
- Updating element styles
- Working with remote data and AJAX requests
- Creating simple tweens and animations

Introduction

There are a number of common and widely used JavaScript frameworks that web professionals use and implement, some of which translate very well into the mobile landscape, such as jQuery.

There are, however, a number of considerations while selecting a JS framework to use in your mobile applications. One is the size of the library, which would inevitably add size to your final packaged application.

While having the full product you may be used to using on your web applications also in use within your mobile apps is serendipitous, options exist for smaller libraries that contain many of the features you need, such as CSS selectors, filtering, style detection, and AJAX requests using XmlHttpRequests.

In this chapter, we will look at the XUI JavaScript library and see how we can use it. XUI was written and maintained by the core members of the PhoneGap development team specifically for use in mobile applications, and it removes many of the unnecessary tools and hidden engines that do not apply to modern browsers available on mobile devices.

Built with mobile devices in mind, XUI is incredibly small and lightweight and works across all devices in the mobile landscape. Unlike some other mobile JavaScript libraries available, XUI does not enforce any page structure or layout. It simply works with the DOM created by the developer to manipulate the layout and work with the content.

Getting started with XUI

Before we can continue with the recipes in this chapter, we must download a copy of the XUI library.

How to do it...

Perform the following steps for the prerequisites to begin with the recipes:

1. Visit `https://github.com/xui/xui`.

   You can either clone the project or click on the **Download ZIP** button, as shown in this screenshot:

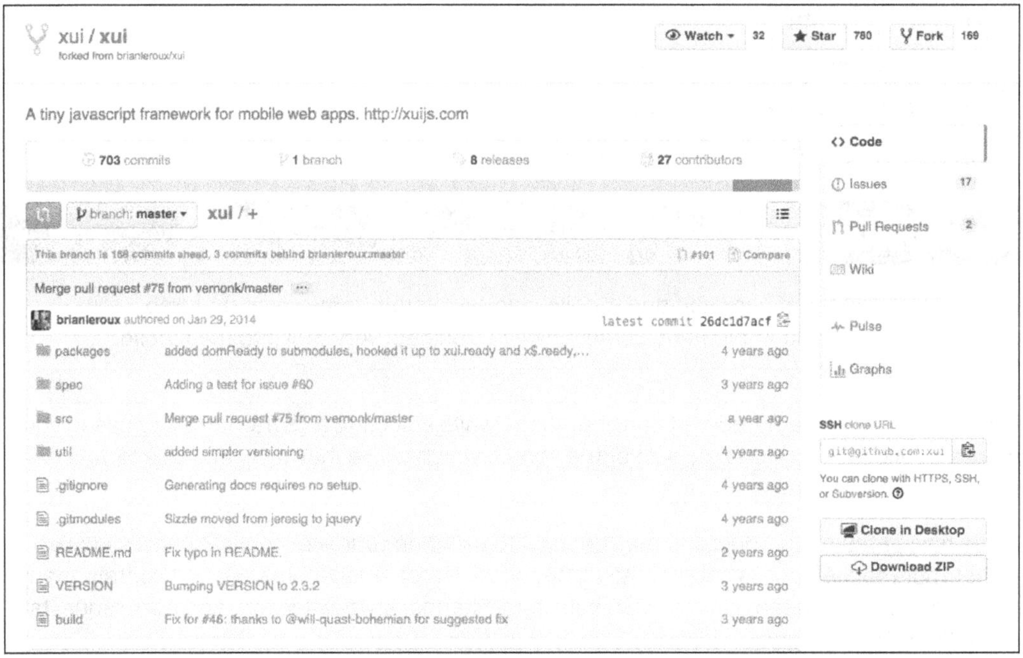

2. XUI can be built to suit your needs. You can either build the library yourself or use the provided `xui.js` for simplicity.
3. We can now proceed with the recipes.

Learning the basics of the library

When creating applications that focus as heavily on user interactions as mobile apps do, we want to be able to easily update and manage the underlying HTML and data collections.

How to do it...

We'll make use of XUI's simple but powerful DOM traversal methods and the ability to extend the library functionality to custom code:

1. Create the basic layout for the HTML page.
2. Create a `div` element within the page with some text. In this case, we're going to create a **Hello World** sample text. Set the `id` attribute for this element to `content`.
3. Include a new `script` tag in the `head` tag of your document, and reference the XUI library within your project directory:

```
<!DOCTYPE HTML>
<html>

<head>
  <meta name="viewport" content="user-scalable=no,
     initial-scale=1, maximum-scale=1,
     minimum-scale=1, width=device-width;" />
  <meta http-equiv="Content-type" content="text/html;
charset=utf-8">
  <title>XUI</title>
  <script type="text/javascript" src="xui.js"></script>

</head>

<body>

  <div id="content">Hello World</div>

</body>
```

4. Before the closing `head` tag, we include a new `script` tag block, inside which we'll place the `onLoad` method, and an XUI `on` event to execute the method once the DOM has fully loaded:

```
<script type="text/javascript">

  x$(window).on('load', onLoad);
```

```
    function onLoad() {

    }

</script>
```

5. Let's use this method to obtain the value of the content element on our page. We can access the XUI library and its methods using the global `x$()` function.
6. Create a new variable called `contentDiv`. We'll pass the `id` of the `div` to the XUI global function to obtain the object reference:

```
function onLoad() {

    var contentDiv = x$('#content');
    console.log(contentDiv);
    console.log(contentDiv.length);
    console.log(contentDiv[0].outerText);

}
```

7. If we run the page in a browser and open the console view, we can see the returned data written to the console log, like this:

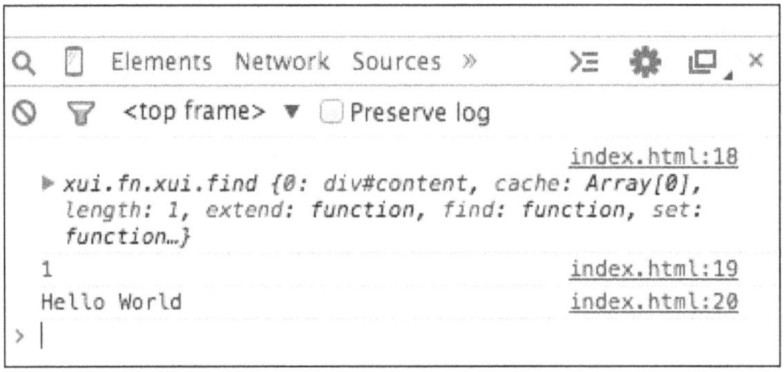

8. The first result is a large structure containing a lot of detailed information about the content `div` element, a sample of which is shown here:

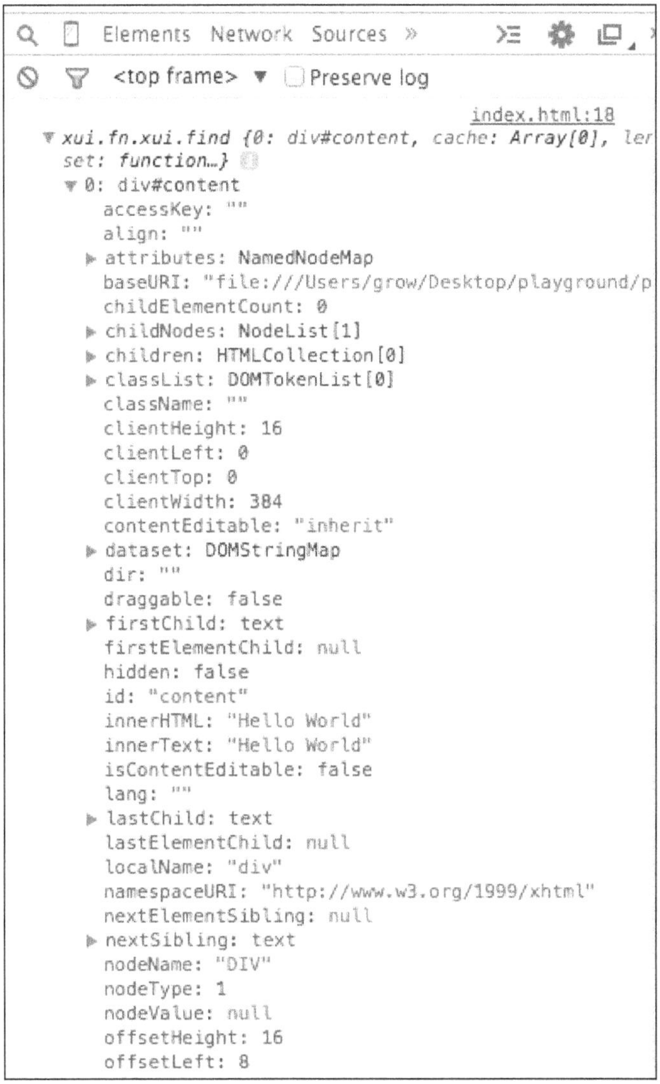

9. The level of details about the returned object is quite extensive, and we can use some of this to programmatically obtain the details about our selected elements.
10. Amend the `div` element to add some more attributes, such as the `style` and `class` attributes, like this:

```
<div id="content"
  style="border: 2px solid #e5e5e5;"
    class="sampleText">
    Hello World
</div>
```

11. Then, amend the `onLoad` method to include the code to obtain the array node containing all the attributes within the selected document element.
12. We'll use this array and loop over the contents to build an HTML string containing the `id` and `value` of each attribute within the `'#content' div` element.
13. Finally, we'll display the generated string after the original `div` element on the page:

```
var contentDiv = x$('#content');

var attributes = contentDiv[0].attributes;

if (attributes.length) {

    var attrMessage = '';

    attrMessage = '<h3>' + contentDiv[0].id + ' ' +
contentDiv[0].localName + ' has ' + attributes.length + '
attributes:</h3>';

    attrMessage += '<ul>';

    for (i = 0; i < attributes.length; i++) {
        attrMessage += '<li>' + attributes[i].localName + ' "'
+ attributes[i].value + '"</li>';
    }

    attrMessage += '</ul>';

    contentDiv.after(attrMessage);
}
```

14. When you run the amended code in your browser, you will see something similar to what is shown here:

Let's now take a look at some of the other useful methods available in the XUI library, which also help form a solid understanding of the basic functionality available:

1. Create a new HTML file, including a reference to the XUI JavaScript library, and an empty `script` tag block before the closing `head` tag.
2. Within the empty script block, define a new `onLoad` method.
3. Add an XUI `on` event handler to run the `onLoad` JavaScript method when the DOM is ready.
4. The `body` of the document will contain two unordered list elements, each with their own `id` attributes, set to `family` and `friends`, respectively. Each list item has a specific `class` attribute that relates to the gender of the individual within the list. Feel free to list your own family and friends.
5. Finally, we'll include a `div` element before the closing `body` tag, with the `id` attribute set to `output`:

```
<!DOCTYPE HTML>
<html>
  <head>
    <meta name="viewport" content="user-scalable=no,
         initial-scale=1, maximum-scale=1,
```

```
        minimum-scale=1, width=device-width;" />
    <meta http-equiv="Content-type"
    content="text/html; charset=utf-8">
    <title>XUI</title>
    <script type="text/javascript"
    src="xui.js"></script>
      <script type="text/javascript">

    x$(window).on('load', onLoad);

    function onLoad() {

    }

    </script>
</head>
<body>

  <ul id="family">
    <li class="male">Ted</li>
    <li class="female">Molly</li>
    <li class="female">Cate</li>
    <li class="female">Jean</li>
   <li class="male">George</li>
  </ul>

  <ul id="friends">
    <li class="male">Steve</li>
    <li class="female">Pip</li>
    <li class="male">Dave</li>
    <li class="male">Scott</li>
  </ul>

  <div id="output"></div>

</body>
</html>
```

6. We'll add some code to the `onLoad` method to filter the various list elements. Firstly, we create an empty `strMessage` variable, in which we'll build our string for the output.
7. We can access all the names within the document list elements, which will provide us with an array of results. We'll store it in the `allNames` variable:

```
function onLoad() {
  var strMessage = '';
```

```
    var allNames = x$('li');

    strMessage += '<p>There are ' + allNames.length
        + ' names in total</p>';
}
```

8. Let's now find elements within specific lists. We can select all the names within our family list by providing a specific element to look into. In this case, we just want to select from the unordered list where the `id` attribute equals `family`.

> Although we explicitly called the `find()` method to locate the list items in this instance, the `x$` namespace is an alias for the `find()` method. As such, we didn't call it in the previous example to obtain all the names in our lists, but we were performing the same function.

9. With can now dig a little deeper and find list items that match a given CSS selector. Here, we will first look for all items within the `allFamily` array that have the `male` class, and then those that have the `female` class. We assign both results to new array variables and add details to our output string:

```
var allFamily = x$('#family').find('li');
var maleFamily = allFamily.has('.male');
var femaleFamily = allFamily.has('.female');

strMessage += '<p>' + allFamily.length + ' family members
are listed:';

strMessage += '<ul><li id="maleFamily">' +
maleFamily.length + ' male</li><li id="femaleFamily">' +
femaleFamily.length + ' female</li></ul></p>';
```

10. Let's do the same to access all the names on our `friends` list.

11. While accessing the class selectors to see whether an element *has* a particular class or not, we can also check whether an element does *not* have a matching CSS selector, as we are doing here for our female friends:

```
var allFriends = x$('#friends').find('li');
var maleFriends = allFriends.has('.male');
var femaleFriends = allFriends.not('.male');

strMessage += '<p>' + allFriends.length + ' friends are
listed:';

strMessage += '<ul><li id="maleFriends">' + maleFriends.length + '
male</li><li id="femaleFriends">' +
femaleFriends.length + ' female</li></ul></p>';
```

12. Finally, we can set the generated string variable in the `div` element for display:

```
x$('#output').html(strMessage);
```

13. Another of XUI's built-in methods allows us to iterate over a collection of elements. Let's combine this into a generic function, which we will use to extend the native XUI library.
14. Create a new variable called `nameFunctions`. It will be an object containing the various functions that we wish to use to extend the library.
15. Name the first reference `generateList`, write the function code to iterate over each element of the provided array, and set the resulting string variable in the parent element:

```
var nameFunctions = {
    generateList: function(array) {
        var list = '<ul>';

        array.each(function(element, index, xui) {
            list += element.outerHTML;
        });

        list += '</ul>';

        this.bottom(list);
    }
}
```

16. Finally, revise the `onLoad` method and add four calls to the `getNameList` function, passing to it each array and list `id` attribute:

```
x$('#output').html(strMessage);

xui.extend(nameFunctions);

x$('#maleFamily').generateList(maleFamily);
x$('#femaleFamily').generateList(femaleFamily);
x$('#maleFriends').generateList(maleFriends);
x$('#femaleFriends').generateList(femaleFriends);
```

Upon running the page in the browser, the output will look something like what is shown here:

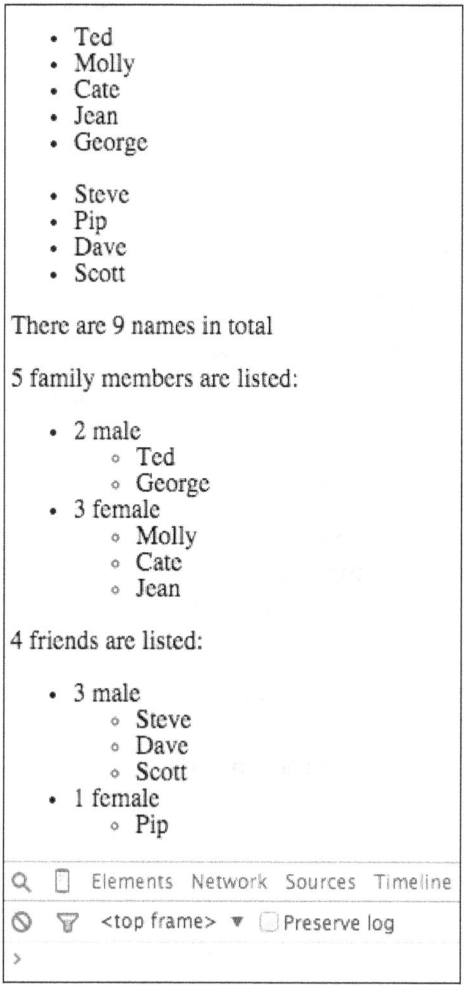

How it works...

In this recipe, we used the find method available in XUI to locate specific elements within our document. We also made use of the `has` and `not` methods to access elements using the CSS selector.

Finally, we extended the functionality offered through XUI by setting our own function in the namespace for easy access to the elements and method calls.

DOM manipulation

Although we may be building our applications in HTML, we still have the ability to alter and manipulate the elements within the document to create a more dynamic application.

How to do it...

We will use the DOM manipulation methods available in the XUI library to read and write content directly within our application:

1. Firstly, create a new PhoneGap project named `dom` by running the following line:

   `phonegap create dom com.myapp.dom dom`

2. Add the devices platform. You can choose to use Android, iOS, or both:

   `cordova platform add ios`

   `cordova platform add android`

3. Open `www/index.html`. Let's clean up the unnecessary elements. Include a reference to the XUI JavaScript library, the Cordova JavaScript file, and an empty `script` tag block before the closing `head` tag.
4. Within the empty `script` block, define a new `onLoad` method.
5. Add an `on` event handler. It will run the `onLoad` JavaScript method when the DOM has fully loaded.
6. The `body` of the document will contain a `div` element with the `id` attribute set to `content`:

```
<!DOCTYPE HTML>
<html>

<head>
    <meta name="viewport" content="user-scalable=no,
         initial-scale=1, maximum-scale=1,
         minimum-scale=1, width=device-width" />
    <meta http-equiv="Content-type" content="text/html;
charset=utf-8">
    <title>XUI</title>
    <script type="text/javascript" src="xui.js"></script>
    <script type="text/javascript"
src="cordova.js"></script>
    <script type="text/javascript">
    x$(window).on('load', onLoad);

    function onLoad() {

    }
```

```
    </script>
</head>

<body onload="onLoad()">

    <div id="content"></div>

</body>

</html>
```

7. Before the `onLoad` method, let's define a new JavaScript object variable. It will act as a container to hold our various HTML strings. We'll access these from `button` click events:

```
var textContent = {
    top: '<div id="top">Top Context</div>',
    inner: '<p>Inner context</p>',
    bottom: '<div id="bottom">Bottom Context</div>',
    after: '<p>DOM Manipulation is easy with XUI!</p>'
};
```

8. Let's now amend the `onLoad` method to include an event listener to verify that the device is ready, at which point the `onDeviceReady` method will fire:

```
function onLoad() {
    document.addEventListener("deviceready",
        onDeviceReady, false);
}
```

9. Include the `onDeviceReady` function. This will make use of the DOM manipulation features and set an `h1` tag before the `content div` element.
10. We'll also set up an event listener to manage button clicks. This will take the `id` attribute of `button`, select the relevant value from our `textContent` object, and insert it into the relevant position in the DOM:

```
function onDeviceReady() {

    x$('#content').before('<h1>XUI DOM Manipulation</h2>');

    x$('button').on('click', function(e) {
        x$('#content').html(this.id, textContent[this.id]);
    });

}
```

11. Back in the HTML, add the following `button` elements at the top of the `body` content. Notice how each button has a specific `id` attribute for referencing the object values:

```
<button id="inner">Inner</button>
<button id="top">Top</button>
<button id="bottom">Bottom</button>
<button id="after">After</button>

<div id="content"></div>
```

12. Finally, let's add some basic styles to the document elements to easily see when they have been added:

```
<style>
  div#content { border: 2px solid #e5e5e5; }
  div#top { background: #ff0000; color: #fff; }
  div#bottom { background: #ffff00; color: #000; }
</style>
```

When you run the application on a device, the output should look something like what is shown here:

How it works...

We created a simple mobile application to see how the XUI library can manipulate the DOM. To set a value within an element, we can use the `html()` function, appended to the XUI element collection.

The `html` function takes two arguments:

- `location`: A `String` that determines the location surrounding the selected element where the DOM manipulation should take place
- `html`: A `String` that contains the HTML markup or elements to be placed in the DOM

The location can be one of the following: `inner`, `outer`, `top`, `bottom`, `remove`, `before`, or `after`.

In our button click handler function, we used the `html` method and passed the `location` string, which we accessed from the `id` attribute of the clicked-on button.

You can also access the locations directly using the shorthand version and simply pass the HTML element or markup, which we did for our `h1` tag entry, like this:

```
x$('#content').before('<h1>XUI DOM Manipulation</h2>');
```

To access the current HTML within an element, we simply need to call the methods without passing an HTML value to set, as follows:

```
// Get the value of the content element
x$('#content').html();
```

Working with touch and gesture events

When working with user interfaces that demand a high level of user interaction or certain processes to be run at certain moments, we need to consider using events and detecting them to execute methods. In this recipe, we will create some functionality that will demonstrate not only the setting of an event but also its removal.

How to do it...

We will use the methods available in the XUI library to control the delegation of event handlers:

1. Firstly, create a new PhoneGap project named `touch` by running the following line:

   ```
   phonegap create touch com.myapp.touch touch
   ```

2. Add the devices platform. You can choose to use Android, iOS, or both:

   cordova platform add ios

   cordova platform add android

3. Add the dialog plugin by running this line:

   cordova plugin add org.apache.cordova.dialogs

4. Open `www/index.html` and clean up the unnecessary elements. Include a reference to the XUI JavaScript library and the Cordova JavaScript file.
5. Add an empty `script` tag block before the closing `head` tag. This will hold our custom PhoneGap code. Within this, define a new `onLoad` function.
6. Add an `on` event handler. It will run the `onLoad` JavaScript method when the DOM has fully loaded.
7. Finally, add a new `button` element within the `body` of the document. Set the `id` attribute to `touchme`:

```
<!DOCTYPE html>

<html>
    <head>
        <meta charset="utf-8" />
        <meta name="format-detection"
content="telephone=no" />
        <meta name="msapplication-tap-highlight"
content="no" />
        <meta name="viewport" content="user-scalable=no,
initial-scale=1, maximum-scale=1, minimum-scale=1,
width=device-width, height=device-height, target-
densitydpi=device-dpi" />
        <script type="text/javascript"
src="xui.js"></script>
        <script type="text/javascript"
src="cordova.js"></script>

        <script type="text/javascript">
            x$(window).on('load', onLoad);

            function onLoad() {

            }
```

```
        </script>

        <title>XUI</title>
    </head>
    <body>
        <button id="touchme">Touch gestures</button>

    </body>
</html>
```

8. Let's add the first of our XUI events. We want to be really sure that the DOM has fully loaded before we manipulate it. The `ready()` method will run once this is the case, so place it within the `onLoad` method.
9. Inside the event, we'll set up our event listener to execute the `onDeviceready` method once the PhoneGap code is ready to run:

```
function onLoad() {
    x$.ready(function() {
        console.log('The DOM is ready to go!');
        document.addEventListener("deviceready",
onDeviceReady, false);
    });
}
```

10. Create the `onDeviceReady` function. This will register a new `touchstart` event to the `button` element:

```
function onDeviceReady() {
    x$('#touchme').on('touchstart', touchConfirmation);
}
```

11. The `touchConfirmation` function will run when the button event has been fired—when it has been touched.
12. Here, we'll display a confirmation notification event from the dialog plugin API to give our users the choice to either keep touching our button or release it:

```
var touchConfirmation = function() {
    navigator.notification.confirm(
        'Do you want to apply a touch gesture again?',
        touchConfirmAction,
        'Touch gesture detected..',
```

```
            'Yes,No'
        )
    };
```

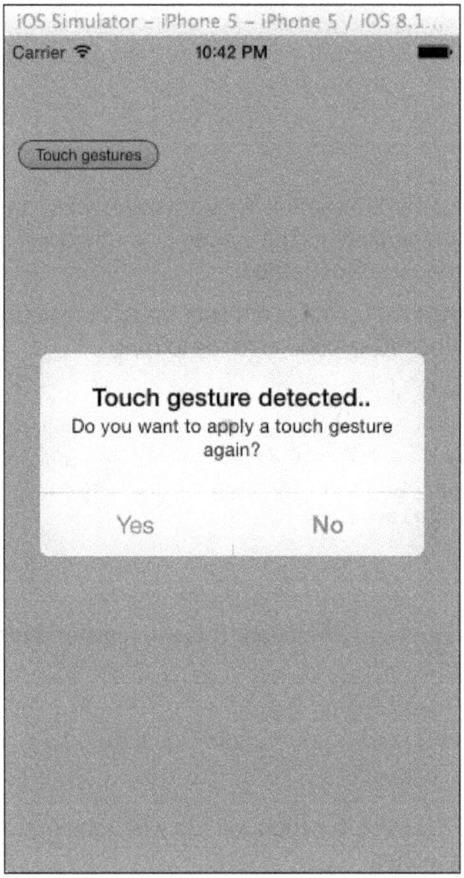

13. Now we need to add the callback function to deal with the selected user option from the confirmation box.
14. If the user selects **Yes**, they can continue to press the button and see the notification window.
15. If they select **No**, we will unregister the event applied to the button, removing the ability to perform any actions when touched.
16. Upon running the application on the browser and selecting **No**, the button value will change. The output will look something like what is shown in this screenshot:

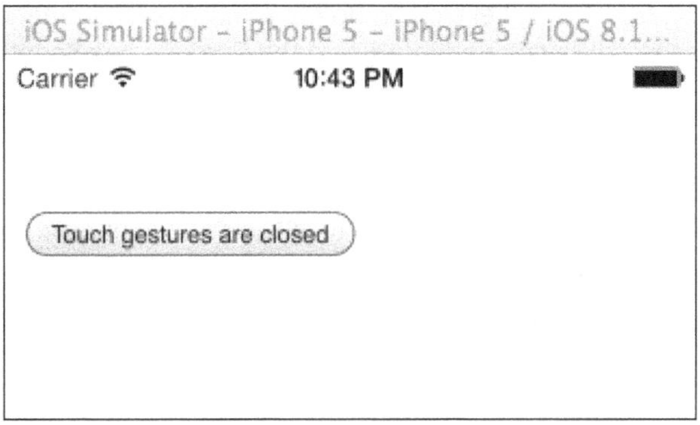

How it works...

In this recipe, we used our first event to detect when the DOM would be ready for us—the `ready()` method.

We also specifically applied a `touchstart` event to a `button` element, which when pressed would alert the user with a notification window.

We then added the functionality of unregistering the specific `touchstart` callback, which would remove the functionality of displaying the notification window.

Updating element styles

As we build our applications and apply a number of styles and properties to define the layout and visual representation of elements, we also need to be able to update and alter the aesthetics and styles to reflect changes or events within the application.

How to do it...

In this recipe, we will read, write, and detect the style properties and class attributes assigned to selected elements using XUI's built-in style functions:

1. Firstly, create a new PhoneGap project named `element` by running this line:

```
phonegap create element com.myapp.element element
```

2. Add the devices platform. You can choose to use Android, iOS, or both:

```
cordova platform add ios
cordova platform add android
```

3. Open `www/index.html`. Let's clean up the unnecessary elements. Include a reference to the XUI JavaScript library and the Cordova JavaScript file.
4. We write a new `script` tag block before the closing `head` tag, which will hold our custom PhoneGap application code. Define an empty `onLoad` function inside the `script` block.
5. Add three `button` elements to the `body` of the document, each with its own individual `id` attribute.
6. Finally, include the `on` event listener to run the `onLoad` function:

```
<!DOCTYPE html>

<html>
    <head>
        <meta charset="utf-8" />
        <meta name="format-detection"
content="telephone=no" />
        <meta name="msapplication-tap-highlight"
content="no" />
        <meta name="viewport" content="user-scalable=no,
initial-scale=1, maximum-scale=1, minimum-scale=1,
width=device-width, height=device-height, target-
densitydpi=device-dpi" />
        <script type="text/javascript"
src="xui.js"></script>
        <script type="text/javascript"
src="cordova.js"></script>

        <script type="text/javascript">
            x$(window).on('load', onLoad);

            function onLoad() {

            }
        </script>

        <title>XUI</title>
    </head>
    <body>
        <button id="one">Button One</button>
        <button id="two">Button Two</button>
        <button id="three">Button Three</button>

    </body>
</html>
```

7. The `onLoad` method will set up the XUI `ready` event handler, which in turn will add the event listener to run the `onDeviceReady` method when the PhoneGap code has been compiled and is ready to run:

```
function onLoad() {
    x$.ready(function() {
        document.addEventListener("deviceready",
            onDeviceReady, false);
    });
}
```

8. When the `onDeviceReady` method executes, the first thing we want to do is register a click handler to all the buttons. We can determine which button was clicked on by reading the `this.id` value that follows a click event.
9. We'll check whether the selected button has a specific CSS class called `active` applied to it. If it does not, we'll add the class to the element.
10. We'll also obtain the `font-size` style attribute applied to the element and send that value to a new method called `resizeFont`. This will increase the `font-size` value by the integer defined in the parameters. In this case, it will add 5 pixels to the value.
11. If the selected button already has the `active` class applied to it, we'll remove the class and reduce the `font-size` style property on the `button` element:

```
function onDeviceReady() {

    x$('button').on('click', function(e) {

        var selectedButton = x$('#' + this.id + '');

        if (!selectedButton.hasClass('active')) {


            selectedButton.addClass('active');
            selectedButton.getStyle('font-size',
                function(property) {
                    resizeFont(selectedButton, property, '+', 5);
                });


        } else {

            selectedButton.removeClass('active');
            selectedButton.getStyle('font-size',
                function(property) {
                    resizeFont(selectedButton, property, '-', 5);
```

```
                    });

                }
            });

        }
```

12. Now let's write the `resizeFont` function. This function will accept four parameters: the `button` element, the value of the `getStyle` property response, whether to increase or decrease the font size, and the amount by which to alter the original property.
13. Once we have calculated the new font size, we'll use the `setStyle` method from XUI to change the style value for the element:

```
function resizeFont(element, property, direction, alterBy)
{
    var sizeType = property.replace(/[0-9]+/, "");
    var fontSize = property.replace(/[^-\d\.]/g, "");
    if ('-' === direction) {
        fontSize = parseInt(fontSize) - alterBy;
    } else {
        fontSize = parseInt(fontSize) + alterBy;
    }
    element.setStyle('font-size', fontSize + sizeType);
}
```

14. To finish off, include some CSS at the bottom of the HTML page to define two states for the buttons—the normal state and the active/selected state:

```
<style>
button {
    background: rgba(100, 100, 100, 0.6);
    color: #fff;
    padding: 5px 10px;
    float: left;
    margin: 5px 5px 0 0;
    border-radius: 2px;
    font-size: 11px;
    font-weight: 600;
    cursor: pointer;
    border: none;
}
button.active {
    background: #EEA839;
```

```
        color: #fff;
        padding: 5px 10px;
        float: left;
        margin: 5px 5px 0 0;
        border-radius: 3px;
        font-size: 11px;
        font-weight: 600;
        cursor: pointer;
        border: none;
    }
    </style>
```

15. If we run the application on a device, the initial screen will look like this, with all three buttons in their normal state:

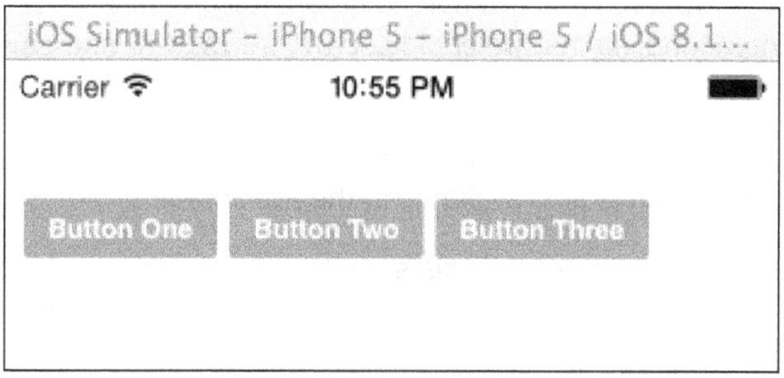

16. If you select a `button` element, you can see that XUI has applied the active style and altered the size of the font. Selecting an active button will remove the styles and revert it to its normal state.

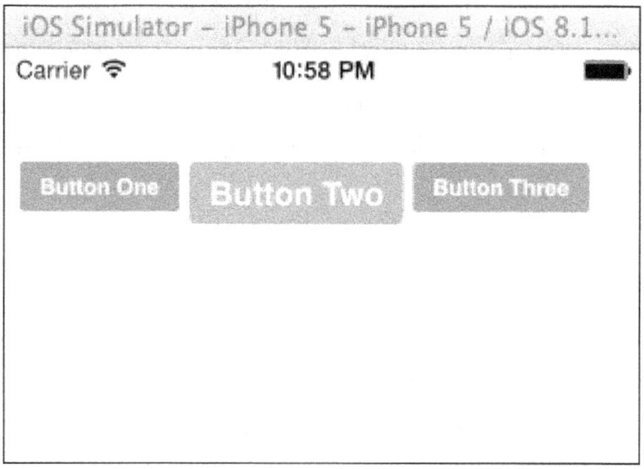

How it works...

In this recipe, we made use of the style manipulation and selector functions in XUI.

After detecting a button `click` event, we were able to check whether the particular button had a specific class, and were able to easily add it to the element if it did not, using the `hasClass` and `addClass` methods, respectively. Similarly, if the class was already applied to the button, we removed it using the `removeClass` method.

We were also able to alter the style properties of our CSS by first accessing it using the `getStyle` method, and then setting the amended property value using `setStyle`.

There's more...

We took a slightly longer route in this example to detect and change the button class attribute. This was intentional so that we could see the `addClass` and `removeClass` methods in action.

We could have removed them both entirely, as well as the `hasClass` method, using the `toggleClass` method, which is also available in the XUI library.

This method adds the specified class it if exists on the selected element, or removes it if it does not, like this:

```
x$('#myButton').toggleClass('active');
```

It's always good to know that you have options!

Working with remote data and AJAX requests

With a vast array of remote servers exposing their services as accessible APIs, you can create some truly dynamic applications by pulling and pushing data to and from external applications and providers.

How to do it...

We will use the XHR method within the XUI library to request data from a remote server, making an asynchronous call to obtain the results from a search:

1. Firstly, create a new PhoneGap project named `ajax` by running the following line:

   `phonegap create ajax com.myapp.ajax ajax`

2. Add the devices platform. You can choose to use Android, iOS, or both:

   cordova platform add ios

   cordova platform add android

3. Open `www/index.html` and clean up the unnecessary elements. Include a reference to the XUI JavaScript library and the Cordova JavaScript file.
4. Create a new `script` tag block before the closing `body` tag with an empty `onLoad` function, which we'll populate shortly.
5. Add an XUI `on` event handler to run the `onLoad` method once the DOM has fully loaded:

```
<!DOCTYPE html>

<html>
    <head>
        <meta charset="utf-8" />
        <meta name="format-detection"
content="telephone=no" />
        <meta name="msapplication-tap-highlight"
content="no" />
        <meta name="viewport" content="user-scalable=no,
initial-scale=1, maximum-scale=1, minimum-scale=1,
width=device-width, height=device-height, target-
densitydpi=device-dpi" />
        <script type="text/javascript"
src="xui.js"></script>
        <script type="text/javascript"
src="cordova.js"></script>

        <script type="text/javascript">
            x$(window).on('load', onLoad);

            function onLoad() {

            }
        </script>

        <title>XUI</title>
    </head>
    <body>

    </body>
</html>
```

6. Within the body of the document, add a text input element with the id attribute set to criteria. Below this, include a new button element with the id attribute set to search.

7. Finally, create a new div block with the id attribute set to response. This will hold our returned data from the remote calls:

```
<body>
    <input value="1" type="number" type="text"
id="criteria" />
    <button id="search">search</button>

    <div id="response"></div>
</body>
```

8. Amend the onLoad method to include a call to the XUI ready event, which will execute when the DOM is ready.

9. Inside of this event handler, add a new event listener to run the onDeviceReady method once the native PhoneGap code is ready, as follows:

```
function onLoad() {
    x$.ready(function() {
        document.addEventListener("deviceready",
            onDeviceReady, false);
    });
}
```

10. Now, let's create the onDeviceReady method. We will apply a touchstart event handler to the button element. This will clear any content that is currently present in the response div element.

11. It will then obtain the value of the criteria input field as specified by the user, and pass that value to a new function:

```
function onDeviceReady() {

    x$('#search').on('touchstart', function(e) {
        x$('#response').html(' ');
        var criteria = x$('#criteria')[0].value;
        generateUser(criteria);
    });

}
```

12. Create a new function called generateUser, which accepts the user input. This will make a simple AJAX call to the randomuser.me API to return *x* entries of a random user.

13. Set the inline callback option to execute a function called `processResults`, which will also contain the response from the request:

```
function generateUser(criteria) {

    x$('#response').xhr(
        'http://api.randomuser.me/?results='+criteria, {
            async: true,
            callback: function() {
                processResults(this.responseText);
            },
            headers: {
                'Mobile': 'true'
            }
        });

}
```

14. The `processResults` method will parse the JSON response from the request. The function will loop over the results so that we can access each entry.

15. We'll use this data to build a simple element collection containing the content of the `tweet` element, which we'll then add at the bottom of the `response div` container:

```
function processResults(response) {
    var response = JSON.parse(response);
    console.log(response);
    var results = response.results;
    for (var i = 0; i < results.length; i++) {
        var data = results[i].user.name.first + ' ' +
results[i].user.name.last;
        var dataDiv = '<div class="tweet">' +
            data + '</div>';
        x$('#response').bottom(dataDiv);
    }
}
```

16. Finally, add some style definitions at the bottom of the document to add some formatting to the returned tweet elements:

```
<style>
.tweet {
    border: 1px solid #999;
    background: #e5e5e5;
    clear: both;
    margin: 5px;
    padding: 5px;
}
</style>
```

Once the user has entered the quantity criteria and pressed the button, the remote call will fire, and the response will look like this:

How it works...

In this recipe, we used the `xhr()` method, available within the XUI JavaScript library, to make a request to a remote web service, the Twitter search API in this case.

Using `xhr()`, we made an `XmlHttpRequest` to the provided URL, which ran an asynchronous call. We passed a callback method called `processResults`. It looped over the returned entries and built an element collection containing the tweet before adding each one at the bottom of the `response` container.

For more information about the `xhr()` method, make sure you check out the official XUI documentation at `https://github.com/xui/xuijs.com/blob/master/views/docs/xhr.ejs`.

Creating simple tweens and animations

Although altering elements in our application's pages is relatively easy, we can further enhance the user's experience by adding some animation to these elements as we change them.

How to do it...

We will use the `tween()` method within the XUI library to alter a property of an image and tween the element to its new position on the screen:

1. Firstly, create a new PhoneGap project named `tweens` by running this line:

   ```
   phonegap create tweens com.myapp.tweens tweens
   ```

2. Add a devices platform. You can choose to use Android, iOS, or both:

   ```
   cordova platform add ios
   cordova platform add android
   ```

3. Open `www/index.html` and clean up the unnecessary elements. Include a reference to the XUI JavaScript library and the Cordova JavaScript file.
4. We add an XUI `ready` event handler to run our code once the DOM has fully loaded:

   ```
   <!DOCTYPE html>

   <html>
       <head>
           <meta charset="utf-8" />
           <meta name="format-detection"
   content="telephone=no" />
           <meta name="msapplication-tap-highlight"
   content="no" />
           <meta name="viewport" content="user-scalable=no,
   initial-scale=1, maximum-scale=1, minimum-scale=1,
   width=device-width, height=device-height, target-
   densitydpi=device-dpi" />
           <script type="text/javascript"
   src="xui.js"></script>
           <script type="text/javascript"
   src="cordova.js"></script>

           <script type="text/javascript">
              x$.ready(function() {


              });
   ```

```
        </script>

        <title>XUI</title>
    </head>
    <body>

    </body>
</html>
```

5. Within the body of the document, add two button elements, one with the id attribute set to up, and the other with the id attribute set to down. Below these, add an empty div tag block with the id attribute set to details.
6. Add a new div tag block with the id attribute set to rocket, inside of which you have to add an image tag with the src attribute referencing the rocket.gif image file:

```
<body>
    <button id="up">UP</button>
    <button id="down">DOWN</button>

    <div id="details"></div>

    <div id="rocket">
        <img src="rocket.gif" width="31" height="72" />
    </div>
</body>
```

7. Now, let's start adding our custom code to the ready() method. Firstly, we'll create two variables that reference the rocket and details div elements.
8. We'll then create a click handler event, applied to the up button. It will obtain the value of the rocket element's bottom CSS property. We can then increment the value by 100 to create a new position for the element.
9. Next, we will send the property to the tween method to animate the rocket element:

```
x$.ready(function() {
    var rocket = x$('#rocket');
    var details = x$('#details');

    x$('#up').on('click', function(e) {

        rocket.getStyle('bottom', function(property) {
            sizeType = property.replace(/[0-9]+/, "");
            topPosition = property.replace(/[^-\d\.]/g, "");
```

```
            topPosition = parseInt(topPosition) + 100;
            newPosition = topPosition + sizeType;
            rocket.tween({
                    bottom: newPosition,
                    duration: 1000
                },
                function() {
                    details.html('bottom: ' + newPosition);
                }
            );
        });

    });

});
```

10. Now, we need to create the click handler event for the down button directly below this, within the `ready()` method, which has the same functionality but decreases the value of the `bottom` property by `100`:

```
x$('#down').on('click', function(e) {

    rocket.getStyle('bottom', function(property) {
        sizeType = property.replace(/[0-9]+/, "");
        topPosition = property.replace(/[^-\d\.]/g, "");
        topPosition = parseInt(topPosition) - 100;
        newPosition = topPosition + sizeType;
        rocket.tween({
                bottom: newPosition,
                duration: 1000
            },
            function() {
                details.html('bottom: ' + newPosition);
            }
        );
    });

});
```

11. Finally, add some style definitions at the bottom of the document to apply the default positions and format for the elements:

```
<style>
  body { background: #0c0440 url(outerspace.jpg) repeat; }
```

```
  div#rocket { bottom: 0px; position: absolute; width:
31px; height: 72px; left: 100px; }
  div#details { color: #ffff00; height: 20px; position:
inherit; top: 20px; right: 20px; float: right; }
</style>
```

12. If we run the application on a device, the initial display will be something like this:

13. If we choose to move the rocket element up, then the result will look like what is shown in the following screenshot:

How it works...

The `tween()` method transforms a CSS property value, in our case the `bottom` property of the `rocket` element. Before we could change the value, we first obtained its current value using XUI's `getStyle()` method. Once we altered the value of the property, we sent it to the `tween()` method. We also sent an optional property to set the duration of the animation in milliseconds.

The `tween()` method accepts two arguments:

- `properties`: An `Object` or `Array` of CSS properties to tween
- `callback`: An optional `Function` that is called when the animation is complete

In our example, we used the optional `callback` argument to update the contents of the `details` element with the new `bottom` property for the `rocket` element.

8

Working with the Ionic Framework

In this chapter, we will cover the following recipes:

- Getting familiar with basics of the library
- Exploring Ionic commands
- Exploring the Ionic framework structure
- Using Cordova

Introduction

This chapter provides basic information about the Ionic framework and how to get started using it. When creating a hybrid mobile application, we often have a problem dealing with the UI component. We want our hybrid app UI to look like a native app. Instead of creating CSS for styling from scratch, it would be easier and faster to use a framework that provides the UI component.

The Ionic framework is an HTML5 framework. It not only provides a native-looking UI component, but also powerful JavaScript framework via Angular JS. The Ionic framework gives a lot of Angular JS directives to make building a hybrid application easier and faster.

The Ionic framework works well with Cordova and PhoneGap. In fact, the Ionic framework **command-line interface** (**CLI**) depends on Cordova CLI to work. Ionic framework provides an easy way to interact with the Cordova plugin by using Cordova.

Getting familiar with basics of the library

There are several ways to get started using Ionic Framework. The easiest way is to utilize Ionic CLI, which is an NPM package. NPM is a dependency manager for Node.js.

How to do it...

Perform the following steps to get started with Ionic framework:

1. Install Ionic CLI by running the following command in the terminal. You may need super user access:

   ```
   npm install -g cordova ionic
   ```

2. If somehow Ionic is not installed, you may require root access:

   ```
   sudo npm install -g cordova ionic
   ```

3. Check your installation by running the following command:

   ```
   ionic -v
   ```

   If you see following output, Ionic has been installed on your machine:

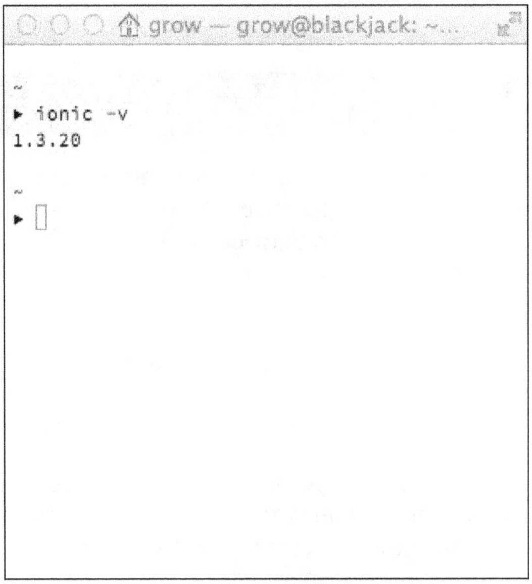

4. Create a new Ionic app project by using a readymade template:

   ```
   ionic start myapp tabs
   ```

5. Run the project by using the following command; you can change `ios` to `android` if you want to emulate on Android instead:

```
cd myapp
ionic platform add ios
ionic build ios
ionic emulate ios
```

6. Your Ionic app is running on the emulator:

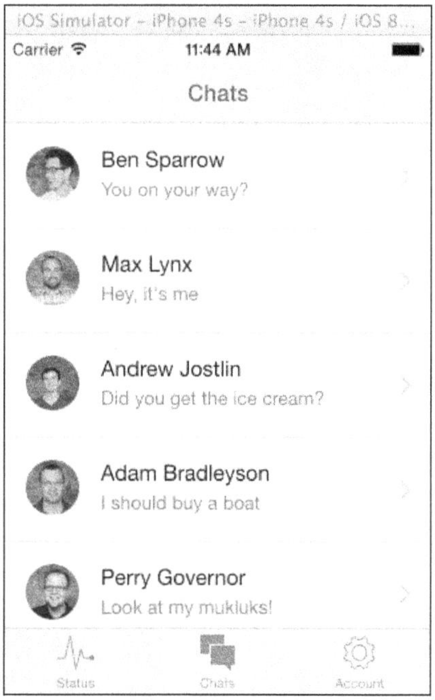

Congratulations! You have successfully created and run an Ionic app.

How it works...

In this recipe, we installed Ionic CLI from the command line/terminal by using `npm`. The `npm` dependency manager downloaded Ionic CLI files and set up the `ionic` command so that we could run the `ionic` command from the command line/terminal.

We also created project by using a readymade template for tabs. This template gives us an easy starting point to build a tabs-based application. Another available template is `sidemenu`. The `sidemenu` template provides a starter template to build an application with side menu navigation. If you prefer to build from scratch, there is a `blank` template. The `blank` template sets up the project directory and provides default HTML and minimal JavaScript code.

We then added a platform to the project by using `ionic platform add <platform>`. This command may be familiar to you if you have worked with other recipes in this book. The `ionic` command will invoke the `cordova` command on certain commands; more will be explained later.

Finally, we built the project and ran the app on the emulator.

Exploring Ionic commands

In this section, we are going to explore Ionic commands. Ionic commands allow us to create, build, and run an Ionic application from the command line or terminal. Working via command line makes application development faster. We don't need to open the specific project, that is, the Android project in IDE, in order to build and run the app.

Ionic commands not only help us by creating, building, and running Ionic-based applications, they also provide other tools to make the development process more enjoyable. By knowing basic Ionic commands, we can improve our workflow while developing hybrid mobile applications.

How to do it...

To get started exploring various Ionic commands, follow these steps:

1. Create a new tabs-based Ionic application by running the following command:

   ```
   ionic start awesomeApp tabs
   ```

2. Change the directory by running the following command:

   ```
   cd awesomeApp
   ```

3. After the project has been generated through Ionic commands, we have to add the `platform` to our application. This can be done by running `ionic platform add <platform>`. To add iOS and Android platforms, run the following commands:

   ```
   ionic platform add ios
   ionic platform add android
   ```

4. For the next step, we will write our application. The default file location is `/www`. We will explore the folder structure later in this chapter.

5. Assuming we don't make any changes to the code, the next step is to run the application. For testing and development purposes, we often prefer to run and debug hybrid application from the browser. Setting up new web servers to test and run applications may be time consuming. Ionic provides a simple command to build a web server that allows us to run and debug applications in the browser:

```
ionic serve
```

6. After the server has been initialized, the `ionic` command will open the browser. Now we can see our application running in the browser:

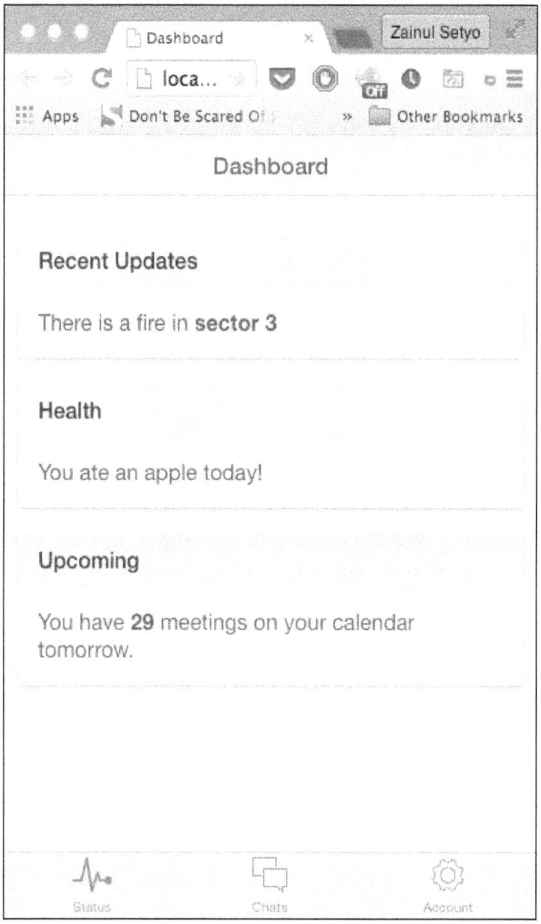

7. The local server generated by the `ionic serve` command uses live reload. This means that whenever we edit and save `*.js` and `*.html`, the browser will automatically refresh the app.

awsomeApp — ionic serve — ionic — node — 65×24
chapter8/02/awsomeApp
▸ ionic serve
Running dev server: http://localhost:8100
Running live reload server: http://localhost:35729
Watching : ['www/**/*', '!www/lib/**/*']
Ionic server commands, enter:
restart or r to restart the client app from the root
goto or g and a url to have the app navigate to the given url
consolelogs or c to enable/disable console log output
serverlogs or s to enable/disable server log output
quit or q to shutdown the server and exit
ionic $ JS changed: www/js/controllers.js
HTML changed: www/index.html

8. The next step is adding plugins to extend the application's capabilities. But we are going to explore plugins and `ngCordova` later in this chapter.
9. After testing it in the browser, we can build the application by using the following commands:

```
ionic build ios
ionic build android
```

10. Next, run the application on the emulator by running:

```
ionic emulate ios
```

11. The application is now running on the emulator:

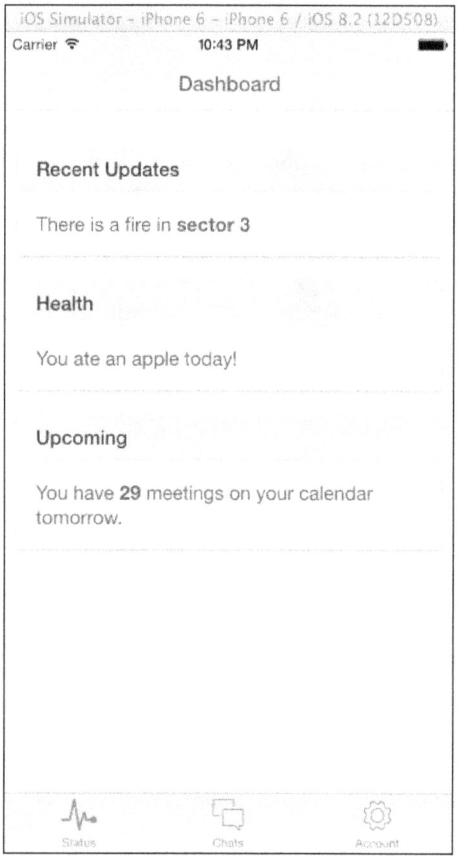

How it works...

First, we created a new tabs-based application by running the `ionic start` command. With this, Ionic downloads project files from the repository to our local directory. This process saves us a lot of time from setting up the Ionic library to work with our Cordova/PhoneGap application.

Then, we ran the `ionic platform` command to add a certain platform to our application. This command downloads the platform-specific project structure. For instance, if iOS platform is added, the Xcode project will be downloaded and placed in our Ionic project.

Once we added the `device` platform, we started the local server for our app by running `ionic serve`. This opens up a browser and runs the application. The server has a built-in live reload feature. It watches our HTML and JavaScript files. Whenever they change, the browser reloads the app instantly.

Finally, we built our Ionic application by running the `ionic build` command. This command builds our application by using a platform-specific project builder. On the Android platform, Ant will be used to build the app. After the build process was completed, we then ran the `ionic` application in the emulator using the `ionic emulate` command.

> We can open official documentation for Ionic framework using the `ionic docs` command. The command will open `http://ionicframework.com/docs/` in the browser.

Exploring the Ionic framework structure

In this section, we are going to explore the Ionic framework project structure. By understanding this structure, we can understand how Ionic framework works. We will start with the basic directory structure, the configuration files, and then the application files.

How to do it...

Basic directory structure

To start exploring the Ionic framework directories and files, follow these steps:

1. Create a new blank Ionic project:

   ```
   ionic start thirdApp blank
   ```

2. Change the directory by running the following command:

   ```
   cd thirdApp
   ```

3. Add the platform to the project:

   ```
   cordova platform add ios
   cordova platform android
   ```

4. Open the project directory. Run the following command if you are using Mac to open the current directory in Finder:

   ```
   open .
   ```

You will see the following directory structure:

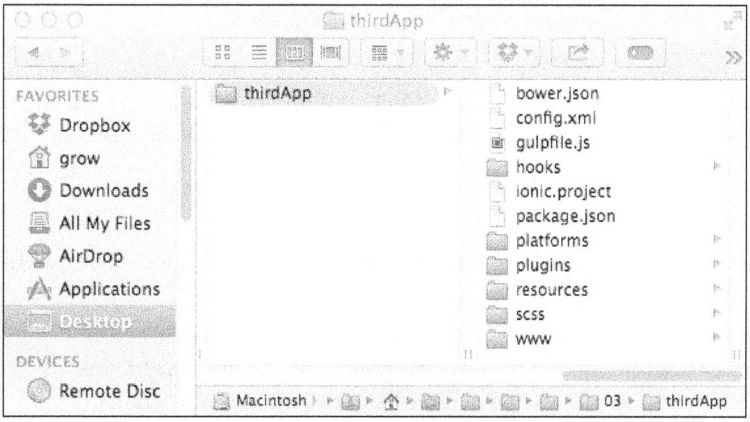

5. You may notice that the project structure is the same with the PhoneGap/Cordova project. Ionic uses the `cordova` CLI behind the scenes. An Ionic project is basically a Cordova project with several additions.

Ionic generates four configuration files: `bower.json`, `gulpfile.js`, `ionic.project`, and `package.json`.

The Cordova project has four directories: `hooks`, `platforms`, `plugins`, `resources`, and `www`.

Configuration and resources

In this recipe, we will discuss configuration files and the resources directory. Configuration files contain the application's configuration: either application-level configuration (`config.xml`, `ionic.project`) or dependency-level configuration (`package.json`, `bower.json`, `gulpfile.js`). The `resources` directory is used to change the application's icon and splash screen. Follow these steps to get a better understanding about Ionic project configuration:

1. The `config.xml` file contains the Cordova project settings. You will see `config.xml` on Cordova, PhoneGap, and Ionic project. Open `config.xml`; the first section of this file is the application's name and description:

```
<?xml version="1.0" encoding="UTF-8" standalone="yes"?>
<widget id="com.ionicframework.thirdapp986055"
version="0.0.1" xmlns="http://www.w3.org/ns/widgets"
xmlns:cdv="http://cordova.apache.org/ns/1.0">
  <name>thirdApp</name>
  <description>
        An Ionic Framework and Cordova project.
    </description>
```

```
    <author email="hi@ionicframework"
href="http://ionicframework.com/">
        Ionic Framework Team
    </author>
```

2. In the next section, you will see default content, which is `index.html`. There will be an `access origin` option to allow the application to access certain URLs.

```
  <content src="index.html"/>
  <access origin="*"/>
```

3. You will then see a bunch of default preferences of the application's behavior:

```
  <preference name="webviewbounce" value="false"/>
  <preference name="UIWebViewBounce" value="false"/>
  <preference name="DisallowOverscroll" value="true"/>
  <preference name="BackupWebStorage" value="none"/>
  <preference name="SplashScreen" value="screen"/>
  <preference name="SplashScreenDelay" value="3000"/>
```

4. There will be a feature named `StatusBar` that shows the option to hide or show the iOS status bar:

```
<feature name="StatusBar">
    <param name="ios-package" value="CDVStatusBar"
onload="true"/>
  </feature>
```

5. The last section of `config.xml` will be platform-specific configuration. The following configuration is used to define the size of each application icon and splash screen images:

```
  <platform name="ios">
    <icon src="resources/ios/icon/icon.png" width="57"
height="57"/>
    <icon src="resources/ios/icon/icon@2x.png" width="114"
height="114"/>
    <icon src="resources/ios/icon/icon-40.png" width="40"
height="40"/>
    <icon src="resources/ios/icon/icon-40@2x.png"
width="80" height="80"/>
    <icon src="resources/ios/icon/icon-50.png" width="50"
height="50"/>
    <icon src="resources/ios/icon/icon-50@2x.png"
width="100" height="100"/>
    <icon src="resources/ios/icon/icon-60.png" width="60"
height="60"/>
    <icon src="resources/ios/icon/icon-60@2x.png"
width="120" height="120"/>
```

```
    <icon src="resources/ios/icon/icon-60@3x.png"
width="180" height="180"/>
    <icon src="resources/ios/icon/icon-72.png" width="72"
height="72"/>
    <icon src="resources/ios/icon/icon-72@2x.png"
width="144" height="144"/>
    <icon src="resources/ios/icon/icon-76.png" width="76"
height="76"/>
    <icon src="resources/ios/icon/icon-76@2x.png"
width="152" height="152"/>
    <icon src="resources/ios/icon/icon-small.png"
width="29" height="29"/>
    <icon src="resources/ios/icon/icon-small@2x.png"
width="58" height="58"/>
    <icon src="resources/ios/icon/icon-small@3x.png"
width="87" height="87"/>
    <splash src="resources/ios/splash/Default-568h@2x~iphone.png"
height="1136" width="640"/>
    <splash src="resources/ios/splash/Default-667h.png"
height="1334" width="750"/>
    <splash src="resources/ios/splash/Default-736h.png"
height="2208" width="1242"/>
    <splash src="resources/ios/splash/Default-Landscape-736h.png"
height="1242" width="2208"/>
    <splash src="resources/ios/splash/Default-Landscape@2x~ipad.
png" height="1536" width="2048"/>
    <splash src="resources/ios/splash/Default-Landscape~ipad.png"
height="768" width="1024"/>
    <splash src="resources/ios/splash/Default-Portrait@2x~ipad.
png" height="2048" width="1536"/>
    <splash src="resources/ios/splash/Default-Portrait~ipad.png"
height="1024" width="768"/>
    <splash src="resources/ios/splash/Default@2x~iphone.png"
height="960" width="640"/>
    <splash src="resources/ios/splash/Default~iphone.png"
height="480" width="320"/>
  </platform>
</widget>
```

> While the `config.xml` file is pretty self-explanatory, you may want to explore other available options for the configuration. You can find a well-documented description of `config.xml` at `https://cordova.apache.org/docs/en/4.0.0/config_ref_index.md.html`.

6. `ionic.project` holds Ionic cloud-related configuration. The Ionic cloud service Ionic View enables us to build applications on the cloud and then test the app using a viewer. The viewer allows us to run the application without having to install the application binary (`.apk` or `.ipa`). Ionic cloud service is beyond the scope of this book.

7. `package.json` is a Node.js configuration that contains project dependencies and other metadata. It may contain the following JSON:

```
{
    "name": "thirdapp",
    "version": "1.0.0",
    "description": "thirdApp: An Ionic project",
    "dependencies": {
        "gulp": "^3.5.6",
        "gulp-sass": "^1.3.3",
        "gulp-concat": "^2.2.0",
        "gulp-minify-css": "^0.3.0",
        "gulp-rename": "^1.2.0"
    },
    "devDependencies": {
        "bower": "^1.3.3",
        "gulp-util": "^2.2.14",
        "shelljs": "^0.3.0"
    },
    "cordovaPlugins": [
        "org.apache.cordova.device",
        "org.apache.cordova.console",
        "com.ionic.keyboard"
    ],
    "cordovaPlatforms": []
}
```

8. `bower.json` is a Bower configuration. Bower is a package manager to manage frontend element dependency, such as CSS and JavaScript. In our application, the only dependency is Ionic:

```
{
    "name": "thirdApp",
    "private": "true",
    "devDependencies": {
        "ionic": "driftyco/ionic-bower#1.0.0-rc.3"
    }
}
```

For more information about Bower, visit `http://bower.io`.

9. `gulpfile.js` contains the configuration for the Gulp task runner. The task runner is particularly helpful to execute repetitive tasks, such as compiling SASS, watching files for live reload feature, and so on. Following are the default tasks for Ionic application:

```
var gulp = require('gulp');
var gutil = require('gulp-util');
var bower = require('bower');
var concat = require('gulp-concat');
var sass = require('gulp-sass');
var minifyCss = require('gulp-minify-css');
var rename = require('gulp-rename');
var sh = require('shelljs');

var paths = {
    sass: ['./scss/**/*.scss']
};

gulp.task('default', ['sass']);

gulp.task('sass', function(done) {
    gulp.src('./scss/ionic.app.scss')
        .pipe(sass({
            errLogToConsole: true
        }))
        .pipe(gulp.dest('./www/css/'))
        .pipe(minifyCss({
            keepSpecialComments: 0
        }))
        .pipe(rename({
            extname: '.min.css'
        }))
        .pipe(gulp.dest('./www/css/'))
        .on('end', done);
});

gulp.task('watch', function() {
    gulp.watch(paths.sass, ['sass']);
});
```

```
gulp.task('install', ['git-check'], function() {
    return bower.commands.install()
        .on('log', function(data) {
            gutil.log('bower', gutil.colors.cyan(data.id),
data.message);
        });
});

gulp.task('git-check', function(done) {
    if (!sh.which('git')) {
        console.log(
            '  ' + gutil.colors.red('Git is not
installed.'),
            '\n  Git, the version control system, is
required to download Ionic.',
            '\n  Download git here:',
gutil.colors.cyan('http://git-scm.com/downloads') + '.',
            '\n  Once git is installed, run \'' +
gutil.colors.cyan('gulp install') + '\' again.'
        );
        process.exit(1);
    }
    done();
});
```

> To get a better understanding about the Gulp task runner and adding new tasks, visit `http://gulpjs.com/`.

10. The resources directory holds icons and splash screens for our application. The easiest way to change the two is by replacing them with the correct resolution images. The resources directory consists of the following contents:

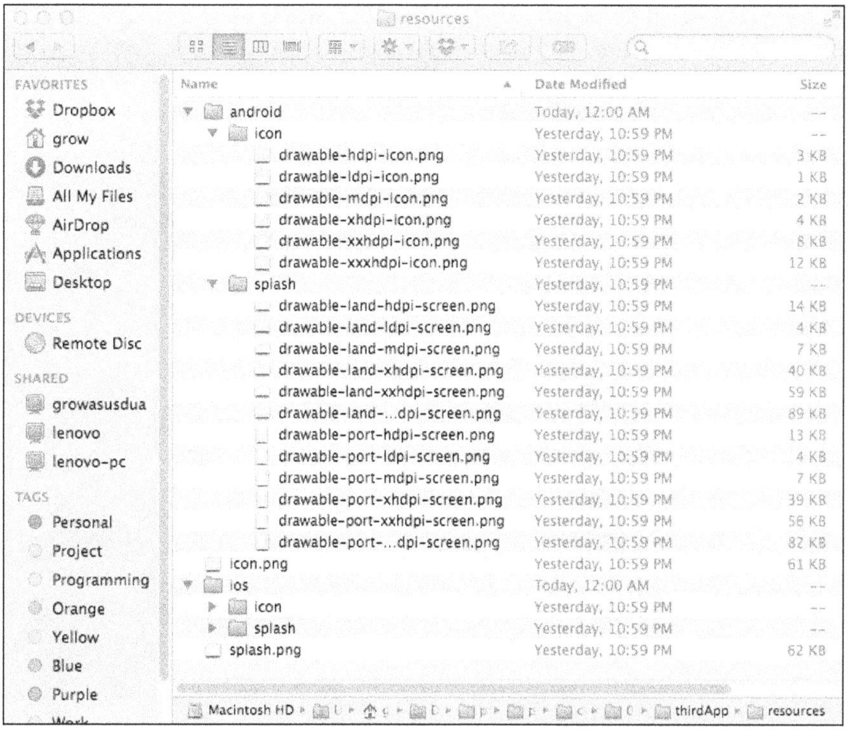

Application files

Application files are located within the www directory. We are going to write our application in the www directory. An Ionic project generates CSS, HTML, and JavaScript files automatically. Follow these steps to dig into the real Ionic framework application files:

1. Open the www directory in the file manager. You will see the following files:

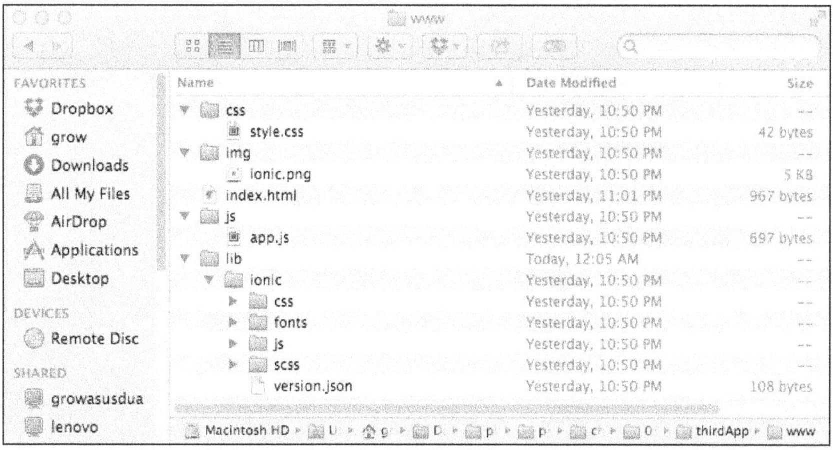

2. The `css`, `img`, and `js` directories are pretty self-explanatory. They contain the application's style sheet, images, and JavaScript respectively, while the `lib` directory holds the application's library. In our application is Ionic itself. The Ionic library consists of a style sheet, fonts, and JavaScript, which produce nice UI elements and components to use.
3. The main file of the Ionic application itself is `index.html`. After opening it, you will get the following content and see a bunch of JavaScript references:

```
<!DOCTYPE html>
<html>

<head>
    <meta charset="utf-8">
    <meta name="viewport" content="initial-scale=1,
maximum-scale=1, user-scalable=no, width=device-width">
    <title></title>
    <link href="lib/ionic/css/ionic.css" rel="stylesheet">
    <link href="css/style.css" rel="stylesheet">
    <!-- IF using Sass (run gulp sass first), then
uncomment below and remove the CSS includes above
    <link href="css/ionic.app.css" rel="stylesheet">
    -->
    <!-- ionic/angularjs js -->
    <script src="lib/ionic/js/ionic.bundle.js"></script>
    <!-- cordova script (this will be a 404 during
development) -->
    <script src="cordova.js"></script>
    <!-- your app's js -->
    <script src="js/app.js"></script>
</head>

<body ng-app="starter">
    <ion-pane>
        <ion-header-bar class="bar-stable">
            <h1 class="title">Ionic Blank Starter</h1>
        </ion-header-bar>
        <ion-content>
        </ion-content>
    </ion-pane>
</body>

</html>
```

4. Add a footer by inserting a new footer component in the app:

```
</ion-content>

    <div class="bar bar-footer bar-balanced">
      <div class="title">Footer</div>
    </div>
</ion-pane>
```

5. Run the `ionic serve` command to see your changes right in the browser:

6. Let's make it more beautiful. Add a new `card` component:

```
<ion-content>


    <div class="list card">
        <div class="item item-avatar">
            <img src="http://placehold.it/50x50">
            <h2>Marty McFly</h2>
            <p>November 05, 1955</p>
        </div>
        <div class="item item-body">
            <img class="full-image"
src="http://placehold.it/50x50">
            <p>
                This is a "Facebook" styled Card. The header is
created from a Thumbnail List item, the content is from a card-
body consisting of an image and paragraph text. The footer
consists of tabs, icons aligned left, within the card-footer.
            </p>
            <p>
                <a href="#" class="subdued">1 Like</a>
                <a href="#" class="subdued">5 Comments</a>
            </p>
        </div>
        <div class="item tabs tabs-secondary tabs-icon-left">
            <a class="tab-item" href="#">
                <i class="icon ion-thumbsup"></i> Like
            </a>
            <a class="tab-item" href="#">
                <i class="icon ion-chatbox"></i> Comment
            </a>
            <a class="tab-item" href="#">
                <i class="icon ion-share"></i> Share
            </a>
        </div>
    </div>

</ion-content>
```

Take a look at the browser:

How it works...

We started with exploring the basic directory structure of an Ionic project. Since this project is built upon a PhoneGap/Cordova project, we saw directories and files of the Ionic project similar to projects in previous chapters. Then we examined the Ionic project files.

We then explored the Ionic configuration files. Ionic framework utilizes third-party tools to improve development workflow. It uses Bower to manage frontend dependencies, GulpJS to manage tasks, and npm to manage application dependency.

Then, we moved to the `www` directory, which contains the Ionic application. We added a footer and card component to the `index.html` file. Adding UI components is a breeze in Ionic. Ionic framework provides several ready-to-use UI components. We will explore UI in a later chapter.

See also

- *Chapter 9, Ionic Framework Development*

Using ngCordova

`ngCordova` is an AngularJS extension on top of the Cordova API. This extension makes building, testing, and deploying Angular JS-based Cordova applications easier. Since Ionic framework is built upon AngularJS, we can utilize `ngCordova` to get easier access to Cordova plugins' API. In this section, we are going to see how to use `ngCordova` along with Ionic framework. We will use the `$cordovaDialogs` call dialog plugin API from the Ionic application. The `$cordovaDialogs` plugin will display a native dialog box for each corresponding platform.

How to do it...

To get started using `ngCordova`, follow these steps:

1. Start by downloading `ngCordova`. Go to `http://ngcordova.com/` and click on the download link. Alternatively, you can download it straight from the GitHub repository by visiting `https://github.com/driftyco/ng-cordova/archive/master.zip`.

2. Then unzip the newly downloaded file; so you have the following files:

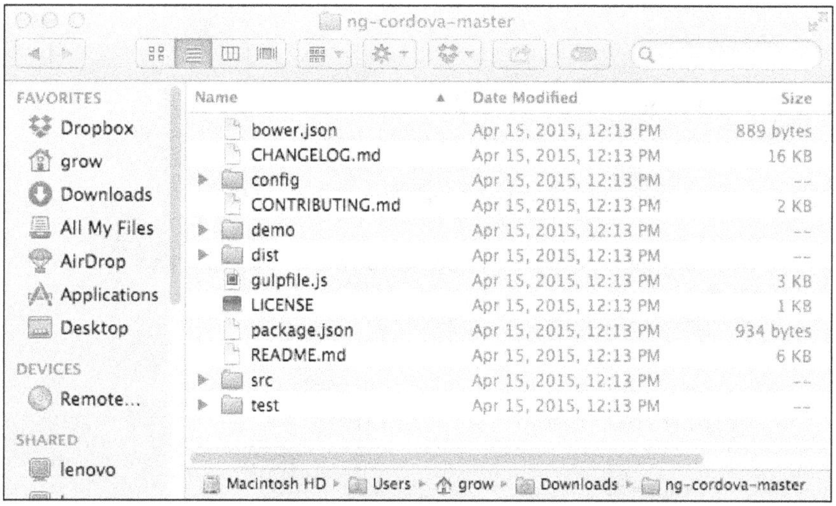

3. The file you need is `dist/ng-cordova.js` or `dist/ng-cordova.min.js`. The `.min` version is just a minified version of the `ngCordova` library. Copy `ng-cordova.js` to be pasted into the Ionic application later.
4. Create a new blank Ionic application project with the name `myngCordova`:

   ```
   ionic start myngCordova blank
   ```

5. Next, add the device platform into the project:

   ```
   ionic platform add ios
   ionic platform add android
   ```

6. Copy `ng-cordova.js` into the `www/js` directory.
7. Open the `www/index.html` file and include `ng-cordova.js` in the JavaScript reference before the `cordova.js` reference and after AngularJS/Ionic:

   ```
   <!DOCTYPE html>
   <html>
     <head>
       <meta charset="utf-8">
       <meta name="viewport" content="initial-scale=1,
   maximum-scale=1, user-scalable=no, width=device-width">
       <title></title>

       <link href="lib/ionic/css/ionic.css" rel="stylesheet">
       <link href="css/style.css" rel="stylesheet">
   ```

```
    <!-- IF using Sass (run gulp sass first), then
uncomment below and remove the CSS includes above
    <link href="css/ionic.app.css" rel="stylesheet">
    -->

    <!-- ionic/angularjs js -->
    <script src="lib/ionic/js/ionic.bundle.js"></script>

    <script src="js/ng-cordova.js"></script>

    <!-- cordova script (this will be a 404 during
development) -->
    <script src="cordova.js"></script>

    <!-- your app's js -->
    <script src="js/app.js"></script>
  </head>
  <body ng-app="starter">

    <ion-pane>
      <ion-header-bar class="bar-stable">
        <h1 class="title">Ionic Blank Starter</h1>
      </ion-header-bar>
      <ion-content>
      </ion-content>
    </ion-pane>
  </body>
</html>
```

8. Now add the dialog plugin:

   ionic plugin add org.apache.cordova.dialogs

9. The preparation is complete, so you can start writing some logics in the Ionic application. Start by injecting the `ngCordova` dependency into the Ionic app. To do this, open `www/js/app.js`. Then add `ngCordova` after the ionic dependency as follows:

```
// Ionic Starter App

// angular.module is a global place for creating,
registering and retrieving Angular modules
// 'starter' is the name of this angular module example
(also set in a <body> attribute in index.html)
// the 2nd parameter is an array of 'requires'
```

```
angular.module('starter', ['ionic', 'ngCordova'])

.run(function($ionicPlatform) {
    $ionicPlatform.ready(function() {
        // Hide the accessory bar by default (remove this
to show the accessory bar above the keyboard
        // for form inputs)
        if (window.cordova &&
window.cordova.plugins.Keyboard) {
   cordova.plugins.Keyboard.hideKeyboardAccessoryBar(true);
        }
        if (window.StatusBar) {
            StatusBar.styleDefault();
        }
    });
})
```

10. Then, add a new controller where you can write the application logic. At the end of `www/js/app.js`, add the following code:

```
.controller('MyController', function($scope,
$cordovaDialogs) {

})
```

11. Define where to use `MyController` in the view. Open `www/index.html`, and then add the `ng-controller="MyController"` attribute within the `ion-pane` tag:

```
<ion-pane ng-controller="MyController">
    <ion-header-bar class="bar-stable">
        <h1 class="title">Ionic Blank Starter</h1>
    </ion-header-bar>
    <ion-content>
    </ion-content>
</ion-pane>
```

12. Now add a new button in the UI along with the `ng-click` event:

```
<ion-pane ng-controller="MyController">
    <ion-header-bar class="bar-stable">
        <h1 class="title">Ionic Blank Starter</h1>
    </ion-header-bar>
    <ion-content class="padding">
        <button ng-click="buttonClicked()"  class="button
button-block button-calm">
            Touch me!!
        </button>
    </ion-content>
</ion-pane>
```

13. When the button is clicked or tapped, it will display a message. Here, JavaScript's `alert` function will not be used. Add the `buttonClicked()` handler in the controller to the call dialog plugin API:

```
.controller('MyController', function($scope,
$cordovaDialogs) {
    $scope.buttonClicked = function() {
        $cordovaDialogs.alert('New message from ngCordova. This
is a native UI component', 'Alert', 'Ok')
            .then(function() {
                // callback success
            });
    };
})
```

14. Build the project and then emulate:

```
ionic build ios
ionic emulate ios
```

15. The application is now running on the emulator. Upon clicking on the button, you will get the following result:

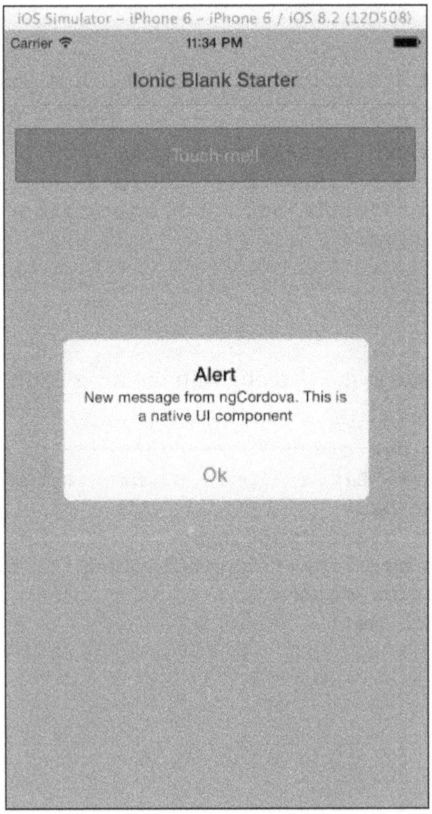

How it works...

We started by downloading the `ngCordova` library from the official repository. Then we created a new blank Ionic project and added device platform. After the project was generated, we placed `ng-cordova.js` into the Ionic project directory which is placed in `www/js`. `ng-cordova.js` reference is placed on `index.html` then.

After the preparation was complete, we injected the `ngCordova` dependency into our AngularJS module so that the `ngCordova` module could be accessed. Then, we created a controller that held our application's logic. Inside `index.html`, we defined which part of the view is controlled by the newly created controller, `MyController`. We then added a button along with the `ng-click` event handler.

Inside `MyController`, we created a new scope to handle the `buttonClicked()` event. Then the `alert()` method of `$cordovaDialogs` was called. This method is used to display alert dialogs using native UI component. We will see different outputs of dialog when running the application on Android and iOS.

> To see all available Cordova plugins on `ngCordova`, visit the official documentation at `http://ngcordova.com/docs/plugins/`.

9

Ionic Framework Development

In this chapter, we will cover these recipes:

- Exploring the UI components
- Creating a layout
- Using Ionic and Angular
- Putting it all together

Introduction

This chapter explains the building of a UI on the Ionic framework. We will start by building a single view application, and move on to a more complex tab-based application. Then, we will cover the usage of AngularJS on the Ionic framework. We will explore routing, the controller, and directives used in Ionic.

Exploring the UI components

In this recipe, we will explore the UI components of the Ionic framework by building a sample UI of settings.

How to do it...

To get started, follow these steps:

1. Create a new blank Ionic application project with the name as `uiComponent`:

   ```
   ionic start uiComponent blank
   ```

2. Change the directory to `uiComponent`:

   ```
   cd uiComponent
   ```

3. Then add the device platforms to the project:

   ```
   ionic platform add ios
   ionic platform add android
   ```

4. We open `www/index.html` and clean `<body>` so that we have the following code:

   ```
   <!DOCTYPE html>
   <html>

   <head>
       <meta charset="utf-8">
       <meta name="viewport" content="initial-scale=1,
   maximum-scale=1, user-scalable=no, width=device-width">
       <title></title>
       <link href="lib/ionic/css/ionic.css" rel="stylesheet">
       <link href="css/style.css" rel="stylesheet">
       <!-- IF using Sass (run gulp sass first), then
   uncomment below and remove the CSS includes above
       <link href="css/ionic.app.css" rel="stylesheet">
       -->
       <!-- ionic/angularjs js -->
       <script src="lib/ionic/js/ionic.bundle.js"></script>
       <!-- cordova script (this will be a 404 during
   development) -->
       <script src="cordova.js"></script>
       <!-- your app's js -->
       <script src="js/app.js"></script>
   </head>

   <body ng-app="starter">
       <div class="bar bar-header bar-positive">
           <h1 class="title">Ionic uiComponent</h1>
       </div>
   </body>

   </html>
   ```

5. Let's add a header and give it a color:

   ```
   <body ng-app="starter">
       <div class="bar bar-header bar-positive">
           <h1 class="title">Ionic uiComponent</h1>
       </div>
   </body>
   ```

6. Then, we will add a footer and give it another color:

```
    <div class="bar bar-footer bar-balanced">
        <div class="title">Footer</div>
    </div>
</body>
```

7. To place our content, we need to add a container for it. We can add the container by adding `div` with the `content` class. As we are using a header in our view, we must add the `has-header` class to the container. Let's add the container below our header and above the footer:

```
<div class="content has-header">

</div>
```

8. Inside the content, we will add a new `list` with items. We create a new container with the `list` class, and then add several list items. The list items will have the `item` class:

```
<div class="content has-header">
    <div class="list">
        <a class="item item-icon-left" href="#">
            <i class="icon ion-email"></i> Check mail
        </a>
        <a class="item item-icon-left item-icon-right"
href="#">
            <i class="icon ion-chatbubble-working"></i> Call Ma
            <i class="icon ion-ios-telephone-outline"></i>
        </a>
        <a class="item item-icon-left" href="#">
            <i class="icon ion-mic-a"></i> Record album
            <span class="item-note">
          Grammy
        </span>
        </a>
        <a class="item item-icon-left" href="#">
            <i class="icon ion-person-stalker"></i> Friends
            <span class="badge badge-assertive">0</span>
        </a>
        <a class="item item-icon-left" href="#">
            <i class="icon ion-person-stalker"></i>
            Friends
            <span class="badge badge-assertive">0</span>
        </a>
    </div>
</div>
```

9. Then, we will add a list divider after the last list item. The list is indicated by the `item-divider` class:

```
<a class="item item-icon-left" href="#">
    <i class="icon ion-person-stalker"></i>
    Friends
    <span class="badge badge-assertive">0</span>
</a>
<div class="item item-divider">
    Activities
</div>
```

10. We will place another list item after the divider, like this:

```
<div class="item item-divider">
    Activities
</div>

<a class="item item-icon-left" href="#">
    <i class="icon ion-flask"></i>
        Breaking Bad
    <span class="item-note">
        Blue, yellow, pink
    </span>
</a>
<a class="item item-icon-left" href="#">
    <i class="icon ion-music-note"></i>
        Music
    <span class="item-note">
        JT
    </span>
</a>
<a class="item item-icon-left" href="#">
    <i class="icon ion-game-controller-b"></i>
        Games
    <span class="item-note">
        Super Mario
    </span>
</a>
```

11. Start the Ionic server to preview the application by running the following line:

```
ionic serve
```

Ionic will open the browser and run the app, as you can see here:

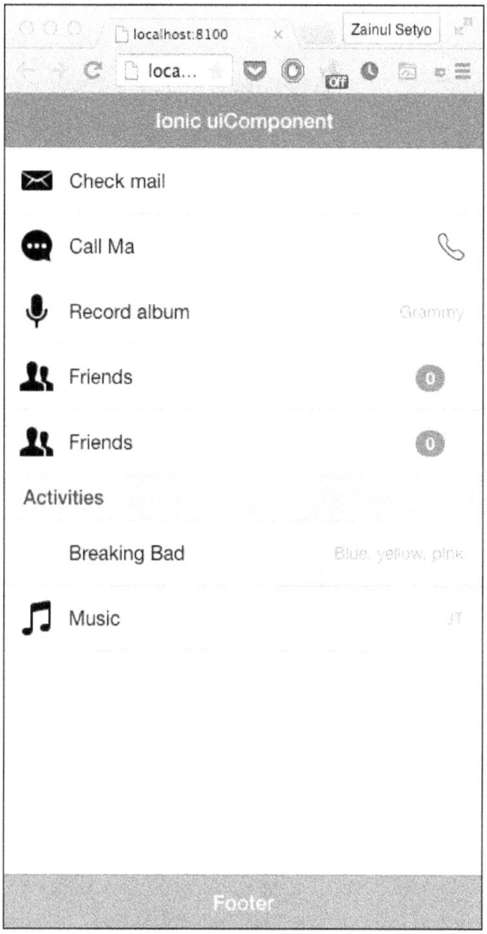

For more information about the UI components, visit `http://ionicframework.com/docs/components/`.

Creating a layout

In the last recipe, we explored the Ionic UI components by creating a single-page application using a blank starter template. In this recipe, we will create a clone of the Instagram mobile app's layout.

How to do it...

To start creating our own version of the Instagram UI, we follow these steps:

1. Create a new tab Ionic application project with the name as `ionSnap`:

   ```
   ionic start ionSnap tabs
   ```

2. Change the directory to `ionSnap`:

   ```
   cd ionSnap
   ```

3. Then add the device platforms to the project:

   ```
   ionic platform add ios
   ionic platform add android
   ```

4. Let's change the tab name. Open `www/templates/tabs.html` and edit each `title` attribute of `ion-tab`:

   ```
   <ion-tabs class="tabs-icon-top tabs-color-active-positive">
       <ion-tab title="Timeline" icon-off="ion-ios-pulse"
   icon-on="ion-ios-pulse-strong" href="#/tab/dash">
           <ion-nav-view name="tab-dash"></ion-nav-view>
       </ion-tab>
       <ion-tab title="Explore" icon-off="ion-ios-search"
   icon-on="ion-ios-search" href="#/tab/chats">
           <ion-nav-view name="tab-chats"></ion-nav-view>
       </ion-tab>
       <ion-tab title="Profile" icon-off="ion-ios-person-outline"
   icon-on="ion-person" href="#/tab/account">
           <ion-nav-view name="tab-account"></ion-nav-view>
       </ion-tab>
   </ion-tabs>
   ```

5. We have to clean our application to start a new tab-based application. Open `www/templates/tab-dash.html` and clean the content so that you will have only the following code:

```
<ion-view view-title="Timeline">
    <ion-content class="padding">
    </ion-content>
</ion-view>
```

6. Then, open `www/templates/tab-chats.html` and clean it up:

```
<ion-view view-title="Explore">
    <ion-content>
    </ion-content>
</ion-view>
```

7. Next, open `www/templates/tab-account.html` and clean that up as well:

```
<ion-view view-title="Profile">
    <ion-content>
    </ion-content>
</ion-view>
```

8. Open `www/js/controllers.js` and delete the methods inside the controllers. So, we will have the following code:

```
angular.module('starter.controllers', [])

.controller('DashCtrl', function($scope) {})

.controller('ChatsCtrl', function($scope, Chats) {

})

.controller('ChatDetailCtrl', function($scope,
$stateParams, Chats) {

})

.controller('AccountCtrl', function($scope) {

});
```

9. We have cleaned up our tabs application. If we run our application now, we will get something like this:

10. In the next step, we will create a layout for the timeline view. Each post of the timeline will be displaying the username, an image, a like button, and a comment button. Open `www/template/tab-dash.html` and add the following `div` list:

```
<ion-view view-title="Timelines">
    <ion-content class="has-header">

        <div class="list card">
            <div class="item item-avatar">
                <img src="http://placehold.it/50x50">
                <h2>Some title</h2>
                <p>November 05, 1955</p>
            </div>
            <div class="item item-body">
                <img class="full-image"
src="http://placehold.it/500x500">
                <p>
                    <a href="#" class="subdued">1 Like</a>
                    <a href="#" class="subdued">5 Comments</a>
                </p>
            </div>
```

```
            <div class="item tabs tabs-secondary tabs-icon-left">
                <a class="tab-item" href="#">
                    <i class="icon ion-heart"></i> Like
                </a>
                <a class="tab-item" href="#">
                    <i class="icon ion-chatbox"></i> Comment
                </a>
                <a class="tab-item" href="#">
                    <i class="icon ion-share"></i> Share
                </a>
            </div>
        </div>

    </ion-content>
</ion-view>
```

Our timeline view will look like this:

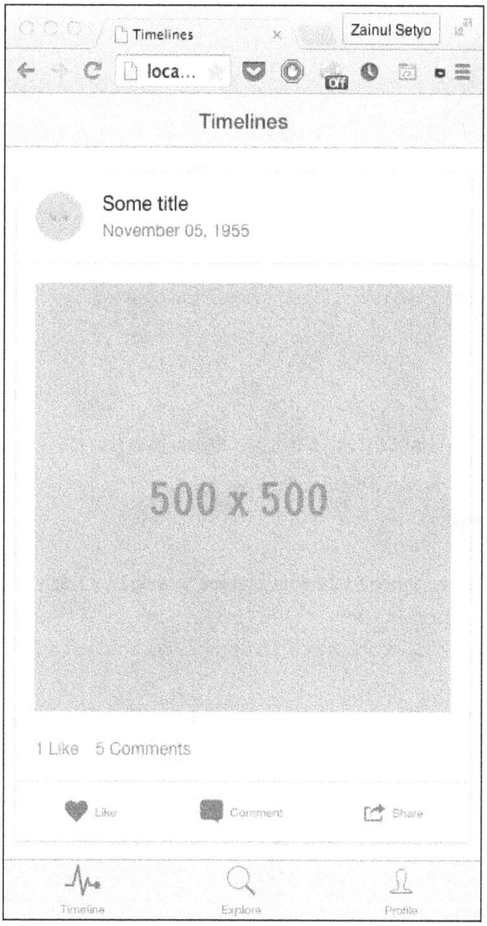

11. Then, we will create an **Explore** page to display photos in a grid view. First, we need to add some styles to our `www/css/styles.css` file:

```
.profile ul {
    list-style-type: none;
}

.imageholder {
    width: 100%;
    height: auto;
    display: block;
    margin-left: auto;
    margin-right: auto;
}

.profile li img {
    float: left;
    border: 5px solid #fff;
    width: 30%;
    height: 10%;
    -webkit-transition: box-shadow 0.5s ease;
       transition: box-shadow 0.5s ease;
}

.profile li img:hover {
    -webkit-box-shadow: 0px 0px 7px rgba(255, 255, 255,
0.9);
    box-shadow: 0px 0px 7px rgba(255, 255, 255, 0.9);
}
```

12. Then we just put the list with the image item, like this:

```
<ion-view view-title="Explore">
    <ion-content>

       <ul class="profile" style="margin-left:5%;">
    <li class="profile">
      <a href="#"><img src="http://placehold.it/50x50"></a>
    </li>
    <li class="profile" style="list-style-type: none;">
      <a href="#"><img src="http://placehold.it/50x50"></a>
    </li>
    <li class="profile" style="list-style-type: none;">
      <a href="#"><img src="http://placehold.it/50x50"></a>
    </li>
```

```
    <li class="profile" style="list-style-type: none;">
      <a href="#"><img src="http://placehold.it/50x50"></a>
    </li>
    <li class="profile" style="list-style-type: none;">
      <a href="#"><img src="http://placehold.it/50x50"></a>
    </li>
    <li class="profile" style="list-style-type: none;">
      <a href="#"><img src="http://placehold.it/50x50"></a>
    </li>
      </ul>

    </ion-content>
</ion-view>
```

Now, our **Explore** page will look like this:

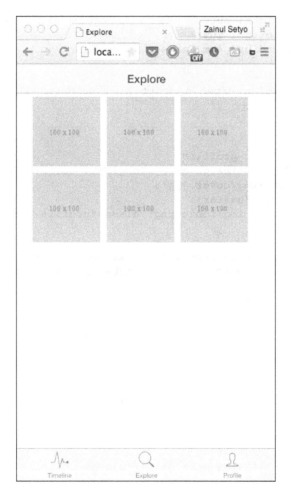

13. Lastly, we will create our profile page. The profile page consists of two parts. The first one is the profile header. It shows user information, such as the username, the profile picture, and the number of posts. The second part is a grid list of pictures uploaded by the user. It's similar to the grid view on the **Explore** page.

14. To add a profile header, open `www/css/style.css` and add the following styles below the existing style:

```
.text-white{
  color:#fff;
}

.profile-pic {
    width: 30%;
    height: auto;
    display: block;
    margin-top: -50%;
    margin-left: auto;
    margin-right: auto;
    margin-bottom: 20%;
    border-radius: 4em 4em 4em / 4em 4em;
}
```

15. Open `www/templates/tab-account.html` and then add the following code inside `ion-content`:

```
<ion-content>

<div class="user-profile"
style="width:100%;heigh:auto;background-color:#fff;float:left;">
    <img src="img/cover.jpg">
    <div class="avatar">
      <img src="img/ionic.png" class="profile-pic">
      <ul>
        <li>
          <p class="text-white text-center" style="margin-top:-
15%;margin-bottom:10%;display:block;">@ionsnap, 6 Pictures</p>
        </li>

      </ul>


    </div>
  </div>
```

16. The second part of the profile page is the grid list of user images. Let's add some pictures under the `profile` header and before the end of the `ion-content` tag:

```
<ul class="profile" style="margin-left:5%;">
  <li class="profile">
    <a href="#"><img src="http://placehold.it/100x100"></a>
```

```
    </li>
    <li class="profile" style="list-style-type: none;">
      <a href="#"><img src="http://placehold.it/100x100"></a>
    </li>
    <li class="profile" style="list-style-type: none;">
      <a href="#"><img src="http://placehold.it/100x100"></a>
    </li>
    <li class="profile" style="list-style-type: none;">
      <a href="#"><img src="http://placehold.it/100x100"></a>
    </li>
    <li class="profile" style="list-style-type: none;">
      <a href="#"><img src="http://placehold.it/100x100"></a>
    </li>
    <li class="profile" style="list-style-type: none;">
      <a href="#"><img src="http://placehold.it/100x100"></a>
    </li>
  </ul>

</ion-content>
```

Our profile page will now look like this:

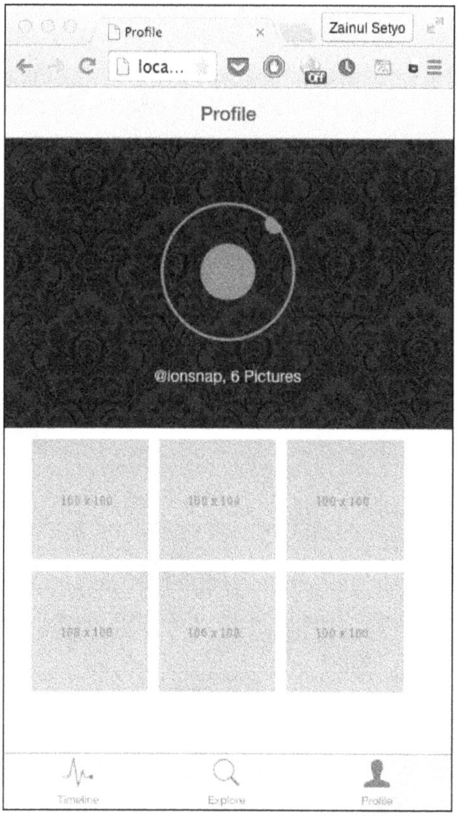

How it works...

We created the app using the tabs template provided by the Ionic team. Using templates can help us get started, especially when starting complex applications, such as a tab-based application. We don't need to create a controller or a separate template, or set up the project. However, the tab-based template needs to be cleaned up. That's why we cleaned up JavaScript and HTML files first.

Then we started creating a layout for the timeline. The timeline consists of multiple photos, and they are displayed in the card view. The timeline layout that we created may seem familiar. It was similar to the Facebook card and the Instagram post. We started by creating a `div` with the `list` and `card` classes. Then we added several `div` with the class item inside the card.

The second layout that we created was the Explore layout. The Explore layout shows a list of pictures in a grid view. We defined a custom style for the `profile` class to display the pictures in a grid view.

The last layout that we created was the profile page. We added a few `profile-pic` styles to `style.css` to make a nice profile header. Then we added a list of user images under the profile header. This list is just a copy of the image list in the Explore layout.

> For more information on building layouts using the Ionic framework, visit `http://ionicframework.com/docs/components/`.

Using Ionic and Angular

In the last recipe, we created an Instagram clone layout using tab templates. We created the layout using the components of the Ionic UI. We skipped the use of AngularJS in our app. In this recipe, we are going to explore and take advantage of AngularJS's controllers, router, and models in our Ionic application.

How to do it...

To start taking advantage of AngularJS's features in an Ionic application, follow these steps:

1. We will continue building the `ionSnap` app that we created in the previous recipe. Open the terminal, change the directory to `ionSnap`, and run the Ionic server:

```
cd path/to/ionSnap
ionic serve
```

2. Open `www/js/app.js` and examine it. The `app.js` is the main entry of our app. There is a bunch of configurations, but for now, we will explore the router only. We can configure each state of our app. Each state holds the configuration of the URL, which template is used, and the name of the controller. The router configuration will look like this:

```
.state('tab.dash', {
    url: '/dash',
    views: {
        'tab-dash': {
            templateUrl: 'templates/tab-dash.html',
            controller: 'DashCtrl'
        }
    }
})
```

3. Our timeline page uses the `templates/tab-dash.html` template and the `DashCtrl` controller. Open `www/js/controllers.js`, and you will find `DashCtrl`. Let's add the posts model inside the `DashCtrl` controller:

```
.controller('DashCtrl', function($scope) {
  $scope.posts = [
    {title: 'First Title', date: 'January 05, 2015', image:
'http://placehold.it/500x500', like: 1, comment: 5},
    {title: 'Second Title', date: 'January 04, 2015', image:
'http://placehold.it/510x510', like: 2, comment: 10},
    {title: 'Third Title', date: 'January 03, 2015', image:
'http://placehold.it/400x400', like: 5, comment: 9}
  ];
})
```

4. We have set up our model. Now, we will bind the model to view. To do this, open `www/templates/tab-dash.html` and change the code to the following:

```
<ion-content class="has-header">
    <div ng-repeat="post in posts" class="list card">
        <div class="item item-avatar">
            <img src="http://placehold.it/50x50">
            <h2>{{post.title}}</h2>
            <p>{{post.date}}</p>
        </div>
        <div class="item item-body">
            <img class="full-image" ng-src="{{post.image}}">
            <p>
                <a href="#" class="subdued">{{post.like}}
Like</a>
                <a href="#" class="subdued">{{post.comment}}
Comments</a>
            </p>
        </div>
```

```
            <div class="item tabs tabs-secondary tabs-icon-left">
                <a class="tab-item" href="#">
                    <i class="icon ion-heart"></i> Like
                </a>
                <a class="tab-item" href="#">
                    <i class="icon ion-chatbox"></i> Comment
                </a>
                <a class="tab-item" href="#">
                    <i class="icon ion-share"></i> Share
                </a>
            </div>
        </div>
    </ion-content>
```

5. Run the app. Our timeline will now have three posts:

6. We will add another model to `ChatsCtrl`, which is the controller of the **Explore** page. Open `www/js/controllers.js` and add the following models to `ChatsCtrl`:

```
.controller('ChatsCtrl', function($scope, Chats) {
    $scope.images = [
    {url: 'http://placehold.it/100x100'},
    {url: 'http://placehold.it/101x101'},
    {url: 'http://placehold.it/102x102'},
    {url: 'http://placehold.it/103x103'},
    {url: 'http://placehold.it/104x104'},
    {url: 'http://placehold.it/105x105'}
    ];
})
```

7. Then we bind the model to the `Explore` page. Open `www/templates/tab-chats.html` and modify the list:

```
<ion-view view-title="Explore">
    <ion-content>

  <ul class="profile" style="margin-left:5%;">
    <li ng-repeat="image in images" class="profile">
      <a href="#"><img ng-src="{{image.url}}"></a>
    </li>
  </ul>

    </ion-content>
</ion-view>
```

Our **Explore** page will look like this:

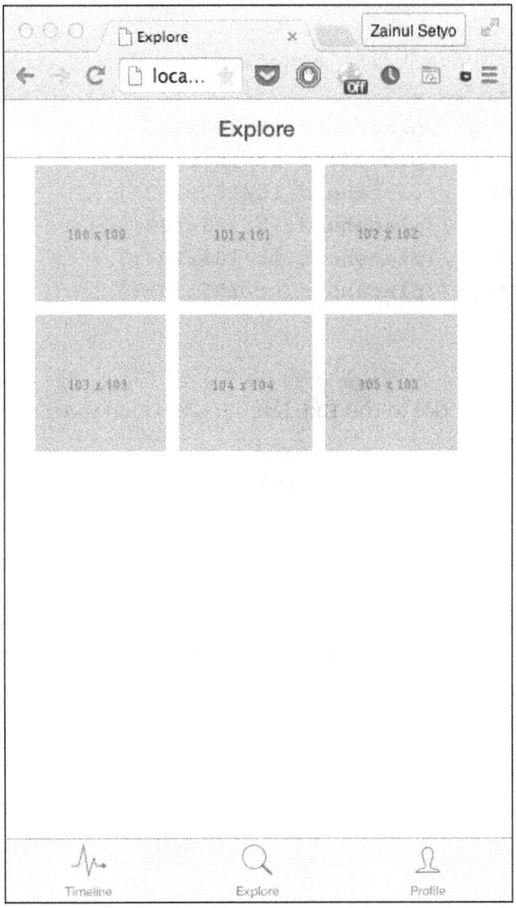

8. Then we will add two models to the profile page: the `userProfile` model and the images model. Open `www/js/controllers.js` and add the following models to `AccountCtrl`:

```
.controller('AccountCtrl', function($scope) {
  $scope.userProfile = {
    username: '@ionsnap',
    avatar: 'img/ionic.png'
    imageCount: 6
  };

  $scope.images = [
    {url: 'http://placehold.it/100x100'},
```

```
    {url: 'http://placehold.it/101x101'},
    {url: 'http://placehold.it/102x102'},
    {url: 'http://placehold.it/103x103'},
    {url: 'http://placehold.it/104x104'},
    {url: 'http://placehold.it/105x105'}
  ];
});
```

9. Next, we bind the models to the view. Open `www/templates/tab-account.html` and change it to the following code:

```
<ion-view view-title="Profile">
    <ion-content>

  <div class="user-profile"
style="width:100%;heigh:auto;background-color:#fff;float:left;">
    <img src="img/cover.jpg">
    <div class="avatar">
      <img ng-src="{{userProfile.avatar}}" class="profile-pic">
      <ul>
        <li>
          <p class="text-white text-center" style="margin-top:-
15%;margin-bottom:10%;display:block;">{{userProfile.username}},
{{userProfile.imageCount}} Pictures</p>
        </li>

      </ul>
    </div>
  </div>

  <ul class="profile" style="margin-left:5%;">
    <li ng-repeat="image in images" class="profile">
      <a href="#"><img ng-src="{{image.url}}"></a>
    </li>
  </ul>

    </ion-content>
</ion-view>
```

The profile page will now have something like what is shown in this screenshot:

How it works...

We started by adding the `posts` model to `DashCtrl`. A model can be assigned using `$scope`. We added the `posts` model to `DashCtrl`. We assigned it using `$scope.posts`. Once the model is assigned to `$scope`, we can access it from the view. Since the `posts` model is an array of objects, we take advantage of `ng-repeat`. We used `ng-repeat="post in posts"`, which means that Angular would loop each of the items on `posts` as `post`. To display the value of the model, we used `{{}}`.

On the profile page, we assigned two models. The `userProfile` is an object that holds information about the user. An object doesn't need `ng-repeat` in order to access the value. We displayed `username` simply by calling `{{userProfile.username}}`.

> For more information about AngularJS, visit `https://docs.angularjs.org/guide`.

Putting it all together

In this recipe, we are going to dig deeper into using AngularJS with the Ionic framework. We are going to add two common features to the existing `ionSnap` app. The first feature is infinite scrolling, which loads more content whenever the user navigates to the bottom of the list. The second feature is pull to refresh, which allows the user to refresh the content by pulling from the top.

How to do it...

To start adding more features into the existing `ionSnap` app, follow these steps:

1. We will continue building the `ionSnap` app we created in an earlier recipe. Open the terminal, change the directory to `ionSnap`, and run the Ionic server:

   ```
   cd path/to/ionSnap
   ionic serve
   ```

2. Then open `www/templates/tab-dash.html`, and add the `ion-infinite-scroll` directive before the `ion-content` close, as shown in this code:

   ```
       <ion-infinite-scroll
           on-infinite="loadMore()"
           distance="1%">
       </ion-infinite-scroll>

   </ion-content>
   ```

3. When the user reaches the end of the list, the `loadMore()` method from the controller will be called. Open `www/js/controller.js` and add the `loadMore` function after `$scope.posts`, like this:

   ```
   $scope.loadMore = function() {

   };
   ```

4. Inside the function, we will prepare dummy data as if it were fetched from the server:

   ```
   $scope.loadMore = function() {

         var dataFetchedFromServer = {
       title: 'Another Title',
       date: 'January 05, 2015',
       image: 'http://placehold.it/500x500',
       like: 1,
       comment: 5
         };
   ```

5. Then we will use JavaScript's `setTimeout` function to produce a delay before appending a new post to the model. After appending a new post to the model, we will send a broadcast message to inform AngularJS that the `infiniteScroll` event is complete:

```
setTimeout(function(){
       // add dataFetchedFromServer to posts model
       $scope.posts.push(dataFetchedFromServer);
       // send broadcast message to indicate the loading process
is complete
       $scope.$broadcast('scroll.infiniteScrollComplete');
}, 3000);
```

Run the app. We will see the loading image at the bottom of the timeline, as shown in this screenshot:

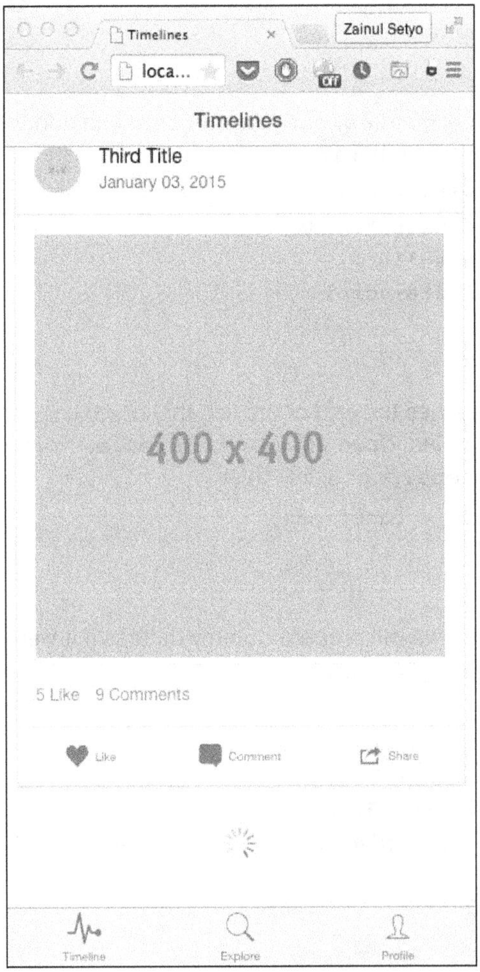

6. The second feature that we will be adding is pull to refresh. We will use the `ion-refresher` directive in our timeline page. Open `www/templates/tab-dash.html` and add the `ion-refresher` directive right after the opening of `ion-content`:

```
<ion-view view-title="Timelines">
    <ion-content class="has-header">
        <ion-refresher
            pulling-text="Pull to refresh..."
            on-refresh="doRefresh()">
        </ion-refresher>
```

7. When pull to refresh is triggered, the directive will call the `doRefresh` method to handle the refresh. Let's create the `doRefresh` method on `DashCtrl` after the `loadMore` method declaration:

```
$scope.doRefresh = function() {

};
```

8. Then we will use `setTimeout` to give a delay before resetting the model. After we are done refreshing the model, we will send a broadcast message to inform AngularJS that the refresh event is complete:

```
$scope.doRefresh = function() {
    var data = [{
        title: 'First Title',
        date: 'January 05, 2015',
        image: 'http://placehold.it/500x500',
        like: 1,
        comment: 5
    }, {
        title: 'Second Title',
        date: 'January 04, 2015',
        image: 'http://placehold.it/510x510',
        like: 2,
        comment: 10
    }, {
        title: 'Third Title',
        date: 'January 03, 2015',
        image: 'http://placehold.it/400x400',
        like: 5,
        comment: 9
    }, {
        title: 'Fourth Title',
        date: 'January 03, 2015',
        image: 'http://placehold.it/400x400',
```

```
            like: 5,
            comment: 9
        }];

        setTimeout(function() {
            $scope.posts = data;
            // Stop the ion-refresher from spinning
            $scope.$broadcast('scroll.refreshComplete');
        }, 3000);

    };
```

9. When we run the `ionSnap` and drag from the header to the bottom, we will see the loading image like this:

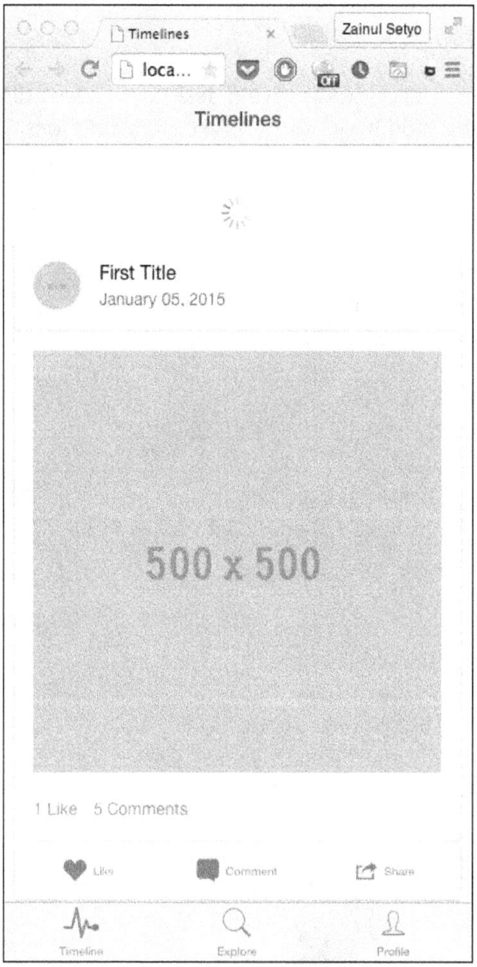

How it works...

We added the infinite scrolling feature to our application. First, we added the `ion-infinite-scroll` directive and set the event to `on-infinite="loadMore()"`. Then we created the implementation of `loadMore()` on `DashCtrl`. We assigned the method using `$scope.loadMore = function() {}`.

Inside the function, instead of getting data from the server, we simulate the process using `setTimeout`. After the process is complete, we send a `$scope.$broadcast('scroll.infiniteScrollComplete')` broadcast message to the framework so that the loading animation disappears.

> For more information about the API of infinite scrolling on the Ionic framework, go to `http://ionicframework.com/docs/api/directive/ionInfiniteScroll/`.

The second feature that we added is pull to refresh. For the first step, we added the `ion-refresher` directive and set the event to `on-refresh="doRefresh()"`. Then we created the implementation of `doRefresh` on `DashCtrl`. Similar to the `loadMore()` method, we assigned the `doRefresh` method using `$scope`. Inside the `doRefresh` method, we initialized fresh data to replace the current model. After the posts model was replaced by the newly initialized fresh data, we sent a `$scope.$broadcast('scroll.infiniteScrollComplete')` broadcast message to stop the animation.

> The complete list of `ion-refresher` APIs, such as changing the loading animation, can be found at `http://ionicframework.com/docs/api/directive/ionRefresher/`.

10
User Interface Development

In this chapter, we will cover these recipes:

- Creating a jQuery Mobile layout
- Persisting data between jQuery Mobile pages
- Using jQuery Mobile ThemeRoller

Introduction

When we develop for mobile devices, we are extremely limited in terms of space, and we have to think differently about creating a user interface or frontend for our application than we would if we were creating it for the "big screen."

Users on mobile devices need information represented clearly and given to them in a format and layout that is easily understandable and recognizable on smaller devices.

In this chapter, we will briefly look into creating a layout for a podcast application, and we will be using the jQuery Mobile framework to do this. With its big, easily identifiable buttons, elements, and interface, it gives us everything that we need to create a layout that almost matches the designs of native apps.

Creating a jQuery Mobile layout

In this recipe, we will be creating a simple podcast application. It will obtain the available shows from a remote XML feed and display them in a list on the main page.

We want the interface and elements to be easily recognizable and simple to use, which jQuery Mobile can easily help us achieve.

Getting ready

Before we start building our application, we need to ensure that we have the jQuery Mobile framework.

Head over to `http://jquerymobile.com/download/` to download the latest code as a `.zip` file.

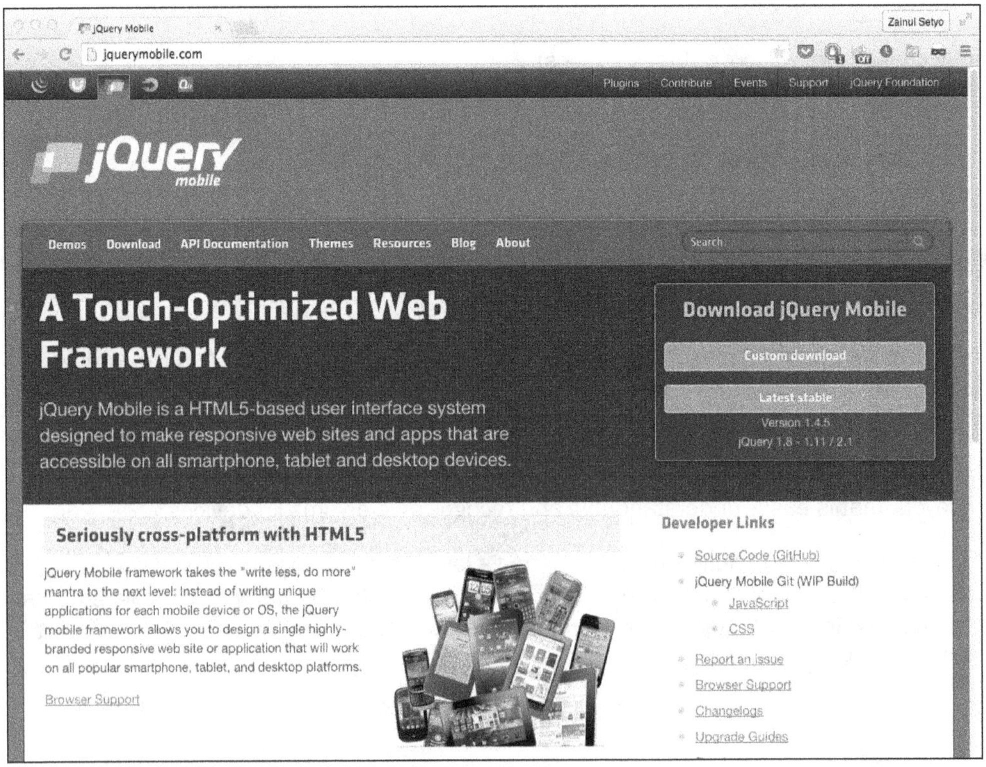

We have options for including the code stored in a **Content Distribution Network** (**CDN**). This means that the files are hosted on a remote server with a fairly high guarantee of their availability at all times, without any server downtime.

Using this method will help to reduce the overall size of our compiled application, as it will mean that our app will not contain the files. They are not gargantuan in size, but when dealing with mobile networks and data use, we want to consider the user by reducing the number of external requests that they may have to make.

We also need to think about connectivity. The users of our application may not have a constant connection to their network, and so we can't rely on remotely stored files. For this reason alone, it is best that we include them in our application.

Once you have downloaded the archive, extract the files to a directory within your application project folder. With this done, we can now get started with creating our jQuery Mobile application.

How to do it...

We will use the jQuery Mobile framework to create the layout and user interface for our mobile application:

1. Create a new `index.html` file in the project folder, which will be your main application page. The head of the document will include a `meta` tag that defines the viewport to assist in defining the structure of the page for use on mobile devices.
2. Include the style sheet reference to the jQuery Mobile CSS file, and two JavaScript references to the jQuery Mobile and jQuery core framework's `.js` files. Let's also include the Cordova JavaScript file within the `head` tag:

```
<!DOCTYPE html>
<html lang="en">
<head>
  <meta name="viewport" content="width=screen.width; user-
scalable=no" />
    <meta http-equiv="Content-type" content="text/html;
charset=utf-8">
    <title></title>
    <link href="jquery.mobile-1.4.5.min.css"
rel="stylesheet" type="text/css" />
    <script src="jquery-1.11.1.min.js"
type="text/javascript"></script>
    <script src="jquery.mobile-1.4.5.min.js"
type="text/javascript"></script>
</head>
<body>

</body>
</html>
```

3. Let's start adding jQuery Mobile-specific code. We'll begin by adding a new page to our application. jQuery Mobile recognizes that certain aspects of the code should be defined as a page by the `data-role` attribute specified within the `div` tag.
4. Each of our pages will have a unique `id` attribute. Here, we'll call our page "home."
5. Create a `div` tag block with `data-role` set to `header`, and a second with `data-role` set to `footer`.
6. We'll also create a new `div` tag with the `data-role` attribute set to `content`, which declares the code within as the content for our page, between the header and footer sections:

```
<body>
    <div data-role="page" id="home">
        <div data-role="header">
            <h1>CFHour Mobile</h1>
        </div>
        <div data-role="content">
            <p>CFHour is the number #1 ColdFusion podcast.</p>
        </div>
        <div data-role="footer">
            <h4>&copy; cfhour.com</h4>
        </div>
    </div>
</body>
```

7. With very little code, we have created a simple application layout. Upon running the application on a device, it will look something like this:

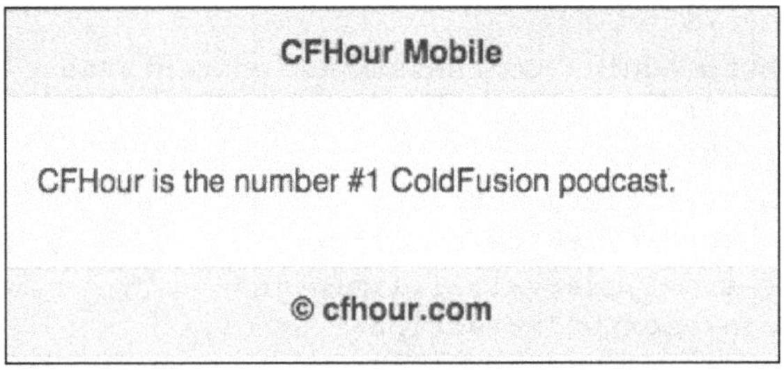

8. As you can see, we have made extensive use of the HTML5 data attributes in our code. jQuery Mobile uses these to define the layout, markup, and behavior of our code.
9. Let's add a second page into our application. Below the current page definition, we'll create a new `div` tag block with the `data-role` attribute set to `page`. Set the `id` attribute for the page to `about` so that jQuery Mobile can differentiate this page from the first.

10. We've also included the header, content, and footer sections, as these need to be defined for each page:

```
<div data-role="page" id="about">
    <div data-role="header">
        <h1>About CFHour</h1>
    </div>
    <div data-role="content">
        <p>CFHour is a weekly podcast primarily focused on
ColdFusion development, but brings you news and updates
about all things 'web'.</p>
        <p>Join your hosts Dave Ferguson, Scott Stroz and
their producer Matt Gifford for the latest information,
live shows and guest interviews.</p>
        <p>
            <a href="http://www.cfhour.com" data-
role="button">Visit www.cfhour.com</a>
        </p>
    </div>
    <div data-role="footer">
        <h4>&copy; cfhour.com</h4>
    </div>
</div>
```

> If you create an application with multiple pages in one file, jQuery Mobile will display the first page that it encounters, in this case the home page. It is important to remember that the order of content in your application will have an effect on what is rendered.

11. We now have our second page created, but we haven't yet created a way for the user to navigate to it. Add an anchor tag within the header of the first page, home, and set the href attribute to point to the second page by referencing its specific id attribute, about. This creates an internal link, and jQuery will know exactly what to do.

12. We can also make use of the mobile framework to turn the standard link into a button, and we'll add an icon from those included in the library to enhance the user interface:

```
<div data-role="page" id="home">
    <div data-role="header">
        <h1>CFHour Mobile</h1>
```

```
        <a href="#about" data-role="button" data-
icon="info">About</a>
</div>
```

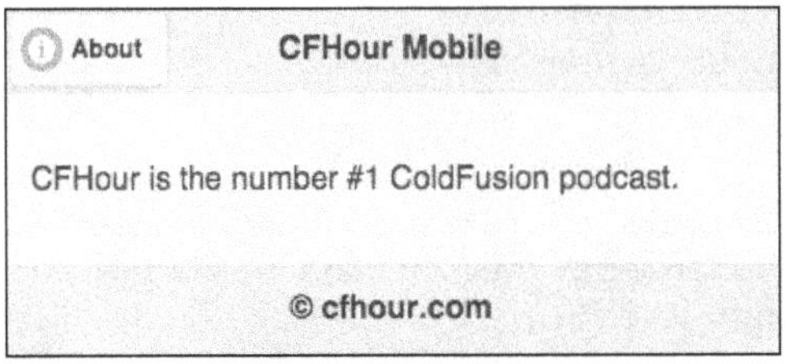

13. We also need to give the user the option to navigate back to the home page. If their mobile device has a back button, they can use it to switch back, but to enhance the user experience, it's best that we provide them with the ability to do so from within the application. Let's amend the **About** page's code to include this:

```
<div data-role="page" id="about">
    <div data-role="header">
        <h1>About CFHour</h1>
        <a href="#" data-rel="back" class="ui-btn-left ui-btn
ui-icon-back ui-btn-icon-notext ui-shadow ui-corner-all"  data-
role="button" role="button">Back</a>
    </div>
```

14. We now have our second page complete, with an automatically generated back button, but as you can see in the screenshot of the application, we also have a lot of empty space below the footer. Let's resolve this and set the footer to sit at the bottom of the screen. Amend the footer `div` tag within each page by adding a new `data` attribute, `data-position`, and set the value to `fixed`, like this:

```
<div data-role="footer" data-position="fixed">
    <h4>&copy; cfhour.com</h4>
</div>
```

15. We now have the header and footer fixed in their positions at the top and bottom of the screen, respectively. This is cleaner for the user, provides a more consistent layout, and also allows the user to scroll to the main content of the page and keep the header and footer locked in position.
16. If we run the application on our device now, the layout will look something like this:

17. When the user loads up the application, we want to provide them with a list of available podcast episodes to listen to. We'll start off by revising the code in the initial home page to include an unordered list:

```
<div data-role="content">
    <p>CFHour is the number #1 ColdFusion podcast.</p>
    <p>Select a show to listen to:</p>
    <ul id="showList" data-role="listview" data-inset="true"></ul>
</div>
```

18. We create a new `script` tag block at the bottom of the file to hold our custom code. To begin with, we'll bind a `pagecreate` jQuery Mobile event to the home page, which will run a new method called `getRemoteFeed`:

```
<script>
    $("#home").bind("pagecreate", function(e) {
        getRemoteFeed();
    })
</script>
```

19. The `getRemoteFeed` method will run a `.get()` jQuery method call to obtain the contents of the podcast feed. When the XML data has been retrieved, we'll loop over each item element (which represents an individual show) and generate a new list item for each one.

    While building your list, you can see that you are adding elements from the feed as data attributes to the list item, as well as the text for the link.

20. After looping through all available items, we append the complete string containing our list items to the `shortList` unordered list element that we created earlier. Finally, we need to refresh the list to allow jQuery Mobile to render the list elements with the correct styles and formatting:

```
var getRemoteFeed = function() {

    $.get(
"http://feeds2.feedburner.com/CfhourColdfusionPodcast",
        {},
        function(data) {

            var listItem = '';

            $(data).find('item').each(function(){


                listItem += '<li data-show-description="'
                + '" data-show-title="'
```

```
                + $(this).find('title').text()
                + '" data-enclosure="'
                + $(this).find("enclosure").attr("url")
                +'" id="'+ $(this).find("guid")
                +'"><h3><a>'
                + $(this).find('title').text()
                + '</a></h3><p>Released: '
                + $(this).find('pubDate').text()
+'</p></li>';
            });

            $('#showList').append(listItem);
            $("#showList").listview("refresh");
        }
    );
};
```

21. Our main application page will now look something like what is shown in this screenshot:

We defined the list to be inset. Had we left this attribute as default, the list would have covered the entire width of the screen.

How it works...

The jQuery Mobile framework gives developers the ability to create application layouts that are responsive to the size of the device screen and provide an entire library of user interface elements that aesthetically match a native application.

We were able to create a layout and define various sections and content simply by using data attributes within the HTML. jQuery Mobile has been built to apply styles and events based on the existence of these data attributes. This means that we only need to write well-structured HTML code—we do not need to worry about diving too deeply into a new language or any development framework that uses the model-view architecture.

Persisting data between jQuery Mobile pages

In this recipe, we will build upon the podcast application built in the previous recipe, extending the functionality and features available for the user.

So far, our podcast application consists of a few simple pages that are independent of each other; that is, any content consumed by a page is used by that page only.

How to do it...

We will use the `localStorage` capabilities to save and retrieve information, causing it to persist across pages:

1. With the list now populated, we need to revise the home method to include a function to capture the tap events on each list item. This will obtain the `title`, `enclosure`, and `description` attribute values from the selected item and set them in the `localStorage` on the device so that we can cause them to persist on the next screen. Then we'll force a page change to a new page called `itemdetail`:

```
$("#home").bind("pagecreate", function(e) {
    getRemoteFeed();
    $('#showList').on('click', 'li', function() {
        localStorage.clear();
        localStorage.setItem("enclosureURL",
            $(this).attr('data-enclosure'));
        localStorage.setItem("showTitle",
            $(this).attr('data-show-title'));
        localStorage.setItem("showDescription",
            $(this).attr('data-show-description'));

        $.mobile.changePage('#itemdetail');
```

```
    });

})
```

2. Let's now create the new `itemdetail` page, which must sit in the code below the initial home page. Remember that the order of the page content in your jQuery Mobile application is important. We will leave the `h1` tag within the header empty, as we'll populate the title with the name of the show from `localStorage`. We have also added another link button inside the header to open up a new page, `showdescription`. This link differs from others that we have implemented as it opens the page in a dialog window. This is because we have specified the `data-rel` attribute as `dialog`.
3. In the contents of the page, we'll define the layout for the audio controls, which will manage the playback of the remote audio file:

```
<div data-role="page" id="itemdetail" data-add-back-btn="true">
    <div data-role="header" data-position="fixed">
        <h1></h1>
        <a href="#showdescription" data-role="button" data-
icon="info" data-rel="dialog" class="ui-btn-right">Description</a>
    </div>
    <div data-role="content">
        <h2 id="showTitle"></h2>
        <a data-role="button" id="playaudio">Play</a>
        <a data-role="button" id="pauseaudio">Pause</a>
        <a data-role="button" id="stopaudio">Stop</a>
        <div class="ui-grid-a">
            <div class="ui-block-a">
                Current: <span id="audio_position">0
sec</span></div>
            <div class="ui-block-b">
                Total: <span id="audio_duration">0</span>
sec</div>
        </div>
    </div>
    <div data-role="footer" data-position="fixed">
        <h4>&copy; cfhour.com</h4>
    </div>
</div>
```

4. Let's now add the JavaScript controls for the audio to the `script` tag. We'll begin by including a `pagebeforeshow` event bound to the `itemdetail` page. This will obtain the `mp3` remote URL and the show title from `localStorage`, and define the playback controls for the audio player:

```
$("#itemdetail").bind("pagebeforeshow", function(e) {
```

```
    var mp3URL = localStorage.getItem("enclosureURL");
    var showTitle = localStorage.getItem("showTitle");

    var audioMedia = null,
        audioTimer = null,
        duration = -1,
        is_paused = false;
```

5. We can now set the show title as the title for the page. Start applying the audio controls:

```
$('#showTitle').html(showTitle);

$("#playaudio").live('tap', function() {
    if (audioMedia === null) {
        $("#audio_duration").html("0");
        $("#audio_position").html("Loading...");
        audioMedia = new Media(mp3URL, onSuccess, onError);
        audioMedia.play();
    } else {
        if (is_paused) {
            is_paused = false;
            audioMedia.play();
        }
    }

    if (audioTimer === null) {
        audioTimer = setInterval(function() {
            audioMedia.getCurrentPosition(
                function(position) {
                    if (position > -1) {
                setAudioPosition(Math.round(position));
                        if (duration <= 0) {
                            duration =
audioMedia.getDuration();
                            if (duration > 0) {
                                duration =
Math.round(duration);

                       $("#audio_duration").html(duration);
                            }
                        }
                    }
                },
                function(error) {
```

```
                        setAudioPosition("Error: " + error);
                    }
                );
            }, 1000);
        }
    });

    function setAudioPosition(position) {
        $("#audio_position").html(position + " sec");
    }
```

6. Include the success and error callback methods:

```
    function onSuccess() {
        setAudioPosition(duration);
        clearInterval(audioTimer);
        audioTimer = null;
        audioMedia = null;
        is_paused = false;
        duration = -1;
    }

    function onError(error) {
        alert('code: ' + error.code + '\n' +
            'message: ' + error.message + '\n');
        clearInterval(audioTimer);
        audioTimer = null;
        audioMedia = null;
        is_paused = false;
        setAudioPosition("0");
    }
```

7. We'll now add the methods used to pause and stop the audio, as well as the event handlers used to detect the `tap` action on the relevant control buttons to run these functions:

```
    function pauseAudio() {
            if (is_paused) return;
            if (audioMedia) {
                is_paused = true;
                audioMedia.pause();
            }
        }

        function stopAudio() {
            if (audioMedia) {
```

```
            audioMedia.stop();
            audioMedia.release();
            audioMedia = null;
        }
        if (audioTimer) {
            clearInterval(audioTimer);
            audioTimer = null;
        }

        is_paused = false;
        duration = 0;
    }

    $("#pauseaudio").live('tap', function() {
        pauseAudio();
    });

    $("#stopaudio").live('tap', function() {
        stopAudio();
    });

});
```

> We won't go into details about the audio functions here, as they are covered in detail in *Chapter 4*, *Working with Audio, Images, and Video*, in the *Playing audio files from the local filesystem or over HTTP* section.

8. When you run the application on the device and select an episode from the list, the show details page will look something like this:

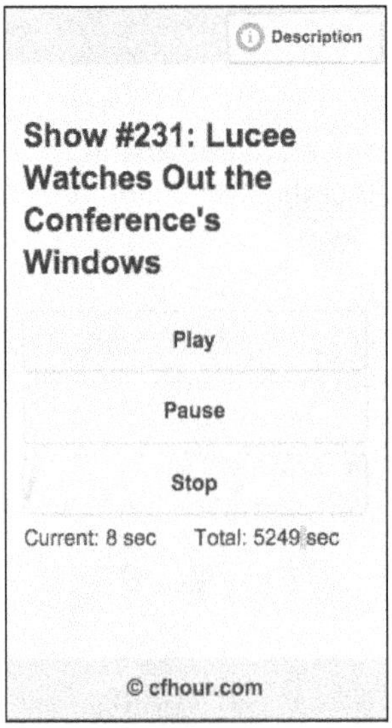

9. Add a new page to the document and set the `id` attribute to `showDescription`. We'll also set the `id` attribute of the content's `div` block to `descriptionContent` and include an empty paragraph tag block, into which we'll insert the content dynamically.
10. Finally, we'll also include a link button that will close the dialog window for the user:

```
<div data-role="page" id="showDescription">
    <div data-role="header">
        <h1>Notes</h1>
    </div>
    <div id="descriptionContent" data-role="content">
        <p></p>
        <a href="#" data-rel="back" data-role="button">Close</a>
    </div>
</div>
```

11. Amend the `.js` file once more and include a new `pagebeforeshow` event binding to the `showDescription` page. This will obtain the `showDescription` key value from `localStorage` and set it in the empty paragraph tags:

```
$("#showdescription").bind("pagebeforeshow", function(e) {
    var description = localStorage.getItem("showDescription");
    $('#descriptionContent p:first').html(description);
});
```

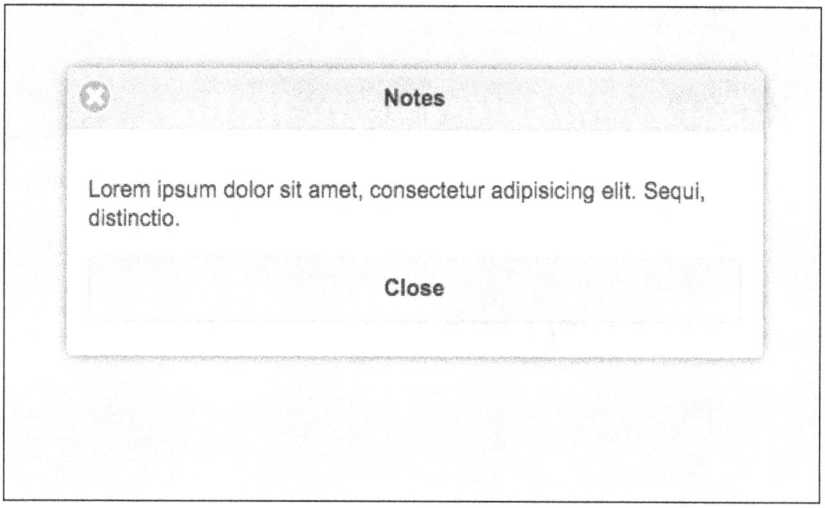

How it works...

We were able to add functionality to the home page to set the touch interaction for each list item. This set the values of the selected individual feed items in the device's `localStorage`, which allowed us to access them using the `localStorage getItem()` method on a new page, independent of the data feed.

We also altered the link to open the `itemDetail` page in a dialog window overlay by setting the `data-rel` attribute for the link itself.

There's more...

You can find out much more about what the jQuery Mobile framework has to offer and how to implement many events, UI elements, and other features in *jQuery Mobile Web Development Essentials*, written by Raymond Camden and Andy Matthews and published by Packt Publishing (`http://www.packtpub.com/jquery-mobile-web-development-essentials/book`).

See also

- The *Caching content using the local storage API* section of *Chapter 3, Filesystems, Storage, and Local Databases*
- The *Playing audio files from the local filesystem or over HTTP* section of *Chapter 4, Working with Audio, Images, and Video*

Using jQuery Mobile ThemeRoller

The jQuery Mobile framework not only provides near-native aesthetics and functionality for page transitions, user interface elements, and page layouts, but also gives us the ability to customize the visual theme of our application. This is managed by changing specific values for the `data-theme` attribute in various elements and containers.

The framework itself ships with five built-in themes, or swatches, alphabetized from A to E.

> For more information on theming your application, read the official jQuery Mobile documentation at `http://jquerymobile.com/demos/1.4.0/docs/api/themes.html`.

Although the provided themes work beautifully and care and consideration have gone into them by the jQuery Mobile team with regards to readability and accessibility, these themes should be considered as only a starting point and not a definitive design for our applications. If you have spent time developing a bespoke native application that interacts with your data and brand, you would also want to make sure that it stands out visually from the crowd and doesn't look like an off-the-shelf theme. Make an impact, and make it individual.

How to do it...

In this recipe, we will explore the features offered by jQuery Mobile ThemeRoller, an online application that allows us to generate our own themes, or swatches, for our mobile application using drag-and-drop interactions:

1. Head over to `http://jquerymobile.com/themeroller/` to begin the creation and customization of your swatches.

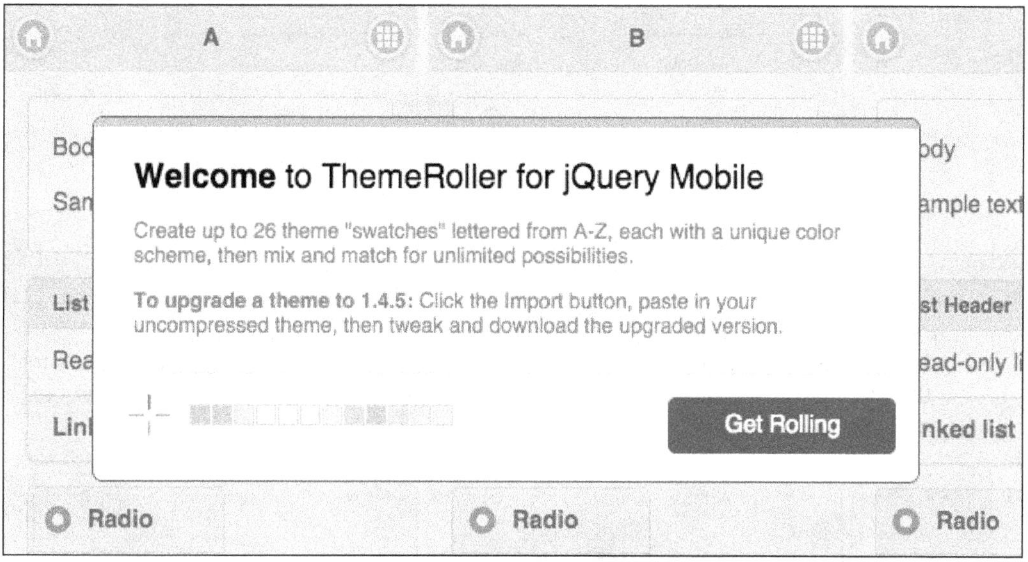

2. The home page will load with a welcome message overlay box, which also reminds you that you have the ability make 26 swatches in total. This should give you plenty of scope to unleash your creative side and explore many color matches and possibilities. Click on **Get Rolling** to dismiss the message.

3. The welcome message recommends that you create a minimum of three swatch variations for your theme. As such, it has rendered three jQuery Mobile page layouts for you to get started with. These are fully interactive so that you can see how your theme looks and feels in a true application layout.

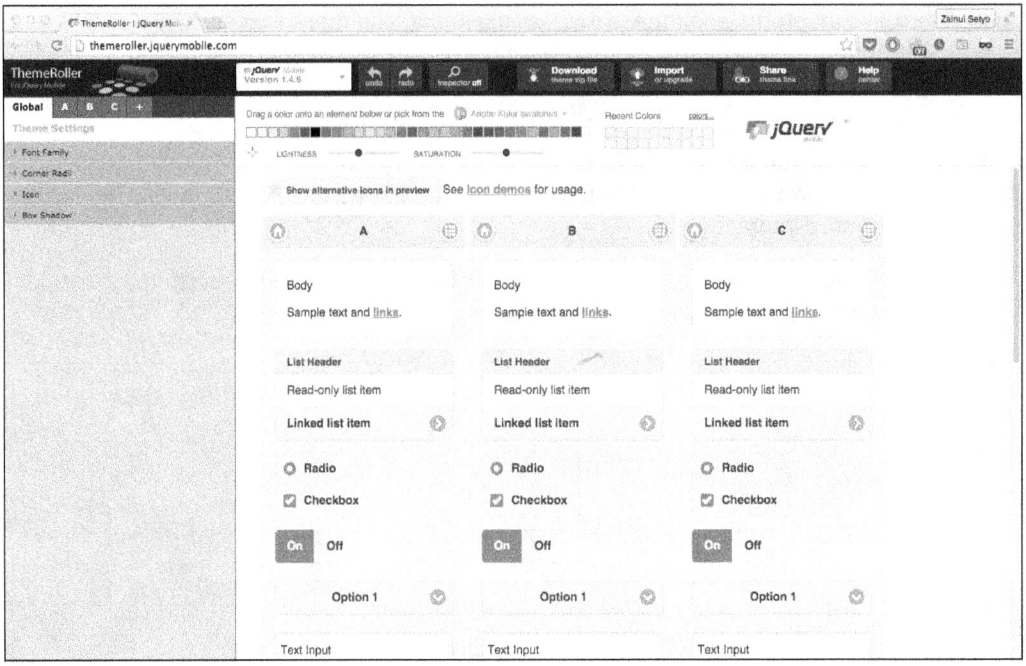

4. To add more than three swatches to your theme, click on the empty layout holder in the main preview panel. This will generate the next swatch and the name in the next alphabetical order.

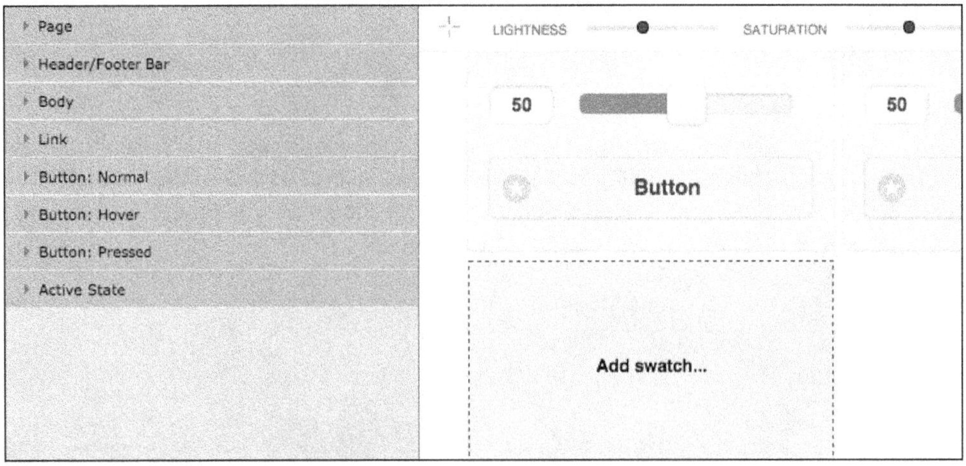

5. The left-hand side of the interface is home to the inspector panel. From here, you can exercise control over the global theme settings, such as the font family, icon color, and corner radii, as shown in the following screenshot. Any changes made to this section will be applied to all the swatches in the preview section.

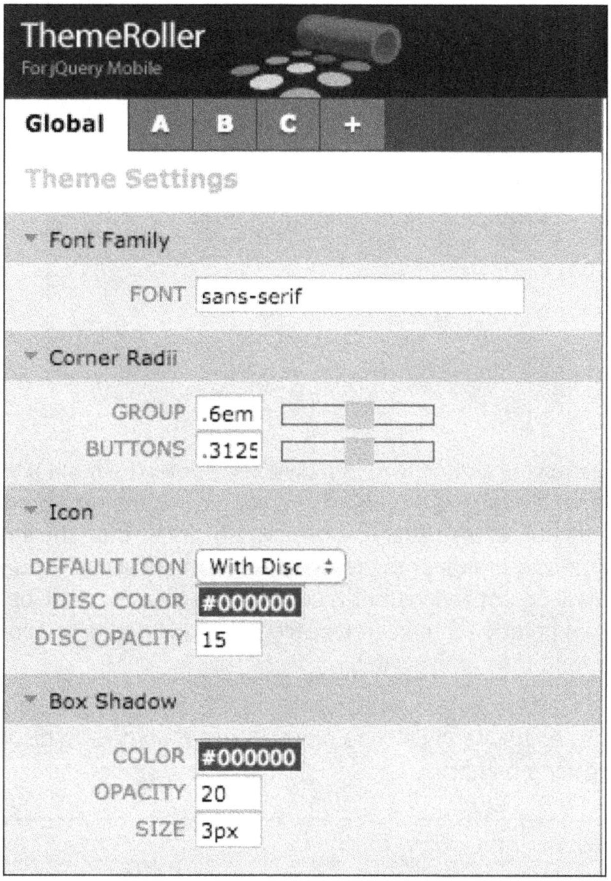

6. Next to the **Global** tab, you also have access to changing theme settings for each individual swatch. This can be accessed by the alphabetical character assigned to each swatch. You can also add a new swatch from here using the **+** button, and it will automatically be placed in the preview section.

7. You can even delete or duplicate a specific swatch from these tabs using the **Delete** link or **Duplicate** link within each swatch tab respectively, as shown here:

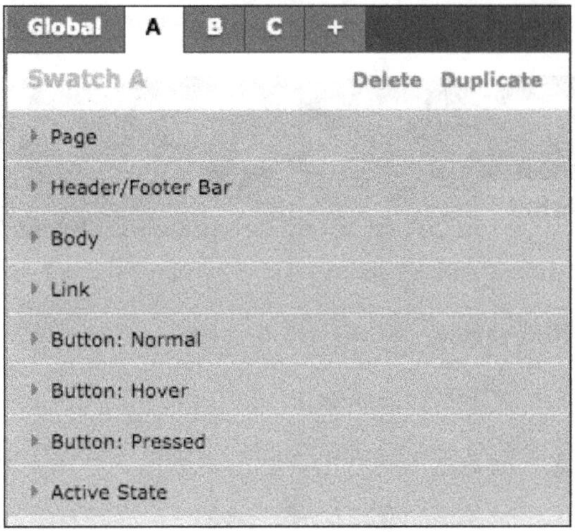

8. Adding colors to your theme is incredibly easy. Above the main preview content, you will see a panel containing a number of color blocks and empty squares. To apply a color to a section of the jQuery Mobile layout, simply drag the selected color block from the panel and drop it onto the UI element on the swatch you wish to update. The change will be applied instantly, and the selected color will be placed in the **Recent Colors** palette for quick reference, should you wish to apply that color elsewhere within the same swatch.
9. You can also adjust the lightness and saturation of each color before applying it to your theme by adjusting the sliders beneath the main palette display, as shown in the following screenshot:

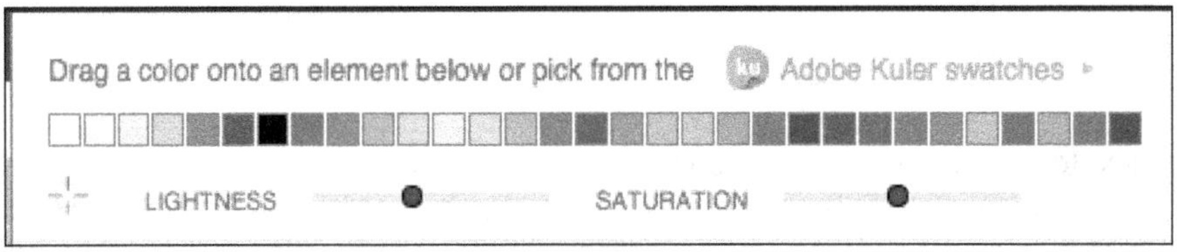

10. If you need some color inspiration or want to see what color schemes other creative professionals have developed, you can access the Adobe Kuler service by clicking on the **Adobe Kuler swatches** link, which will display visual representations of the latest swatches generated through the service. You can also filter for the most popular or selected random themes.

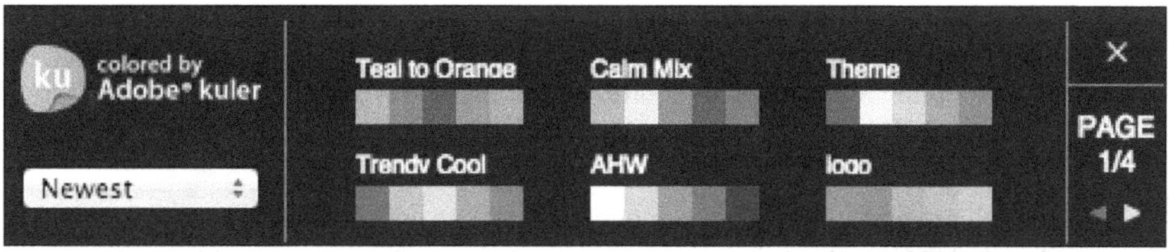

11. Moreover, you can create your own free account with Adobe Kuler (now Adobe Color CC) to start generating and sharing your own swatches. To find out more, visit `http://kuler.adobe.com/`.

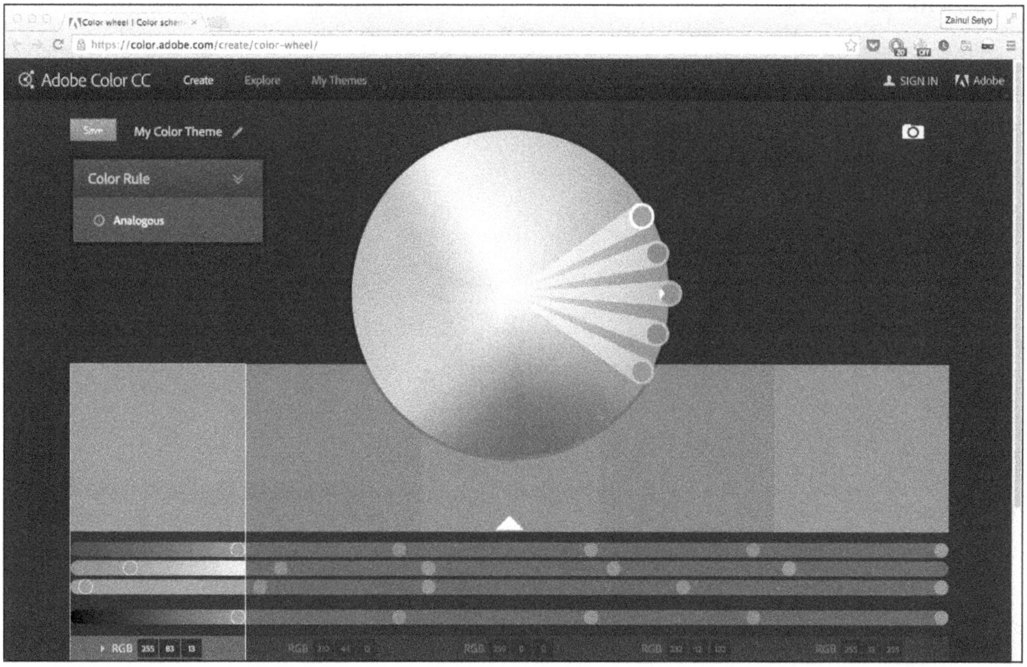

12. The toolbar at the top of the interface gives you instant access to some useful tasks and features, including the **undo** and **redo** options, as shown in this screenshot:

13. If you have an existing theme that you would like to amend or revise, you can import it directly into the interface using the **Import** option. This will generate the correct number of layouts for each swatch in the theme, and apply the styles to each one so that you have an immediate visual representation of the theme.

14. A very useful tool is **Inspector**. Click on it to turn it on or off. When it is turned on, you simply need to hover over a specific part of the layout and click on the selection. This will open up the specific panel and tab in the **Inspector** panel on the left-hand side of the interface. This is great for accessing the portion of the layout that you wish to edit immediately.
15. Once you are happy with your color choices, you can choose to download your completed theme. This will package the relevant files in a `.zip` file, in both compressed and uncompressed formats. Simply import the CSS files into your project location, and include the reference to the style sheet within the `head` tags of your document, like this:

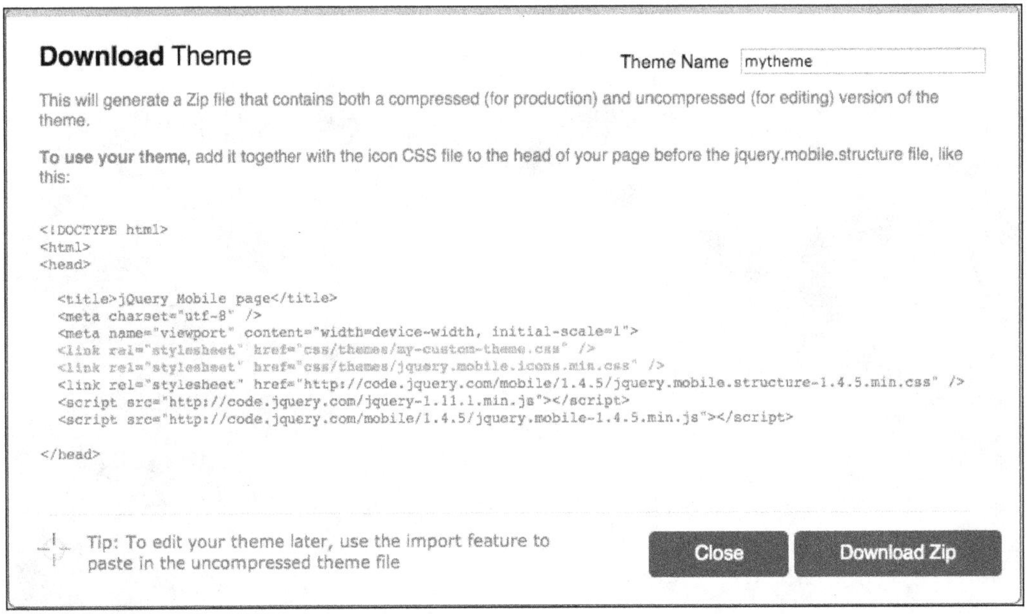

How it works...

Using a simple online interface, we have the ability to create color swatches and themes that better suit our applications' brand while still using the power offered by the jQuery Mobile framework.

There's more...

A framework is a framework, and offers a standard approach to fulfill common design and development tasks. Themes generated by the ThemeRoller application (and indeed the themes that come with the jQuery Mobile framework) still look like every other jQuery Mobile theme—they do share common interface elements, after all.

One thing to remember is that these themes are nothing more than CSS, and as such, you can create truly bespoke layouts and designs with the right amount of skill, time, and patience. Your mobile application need not look like a clone of every other application in the marketplace that uses the same framework, and it can still retain the common user interface elements that help define it as a mobile app.

For some inspiration on what you can achieve with some CSS, take a look at the JQM Gallery site, which showcases some wonderful designs created using the jQuery Mobile framework at `http://www.jqmgallery.com/`. A screenshot of this website is shown here:

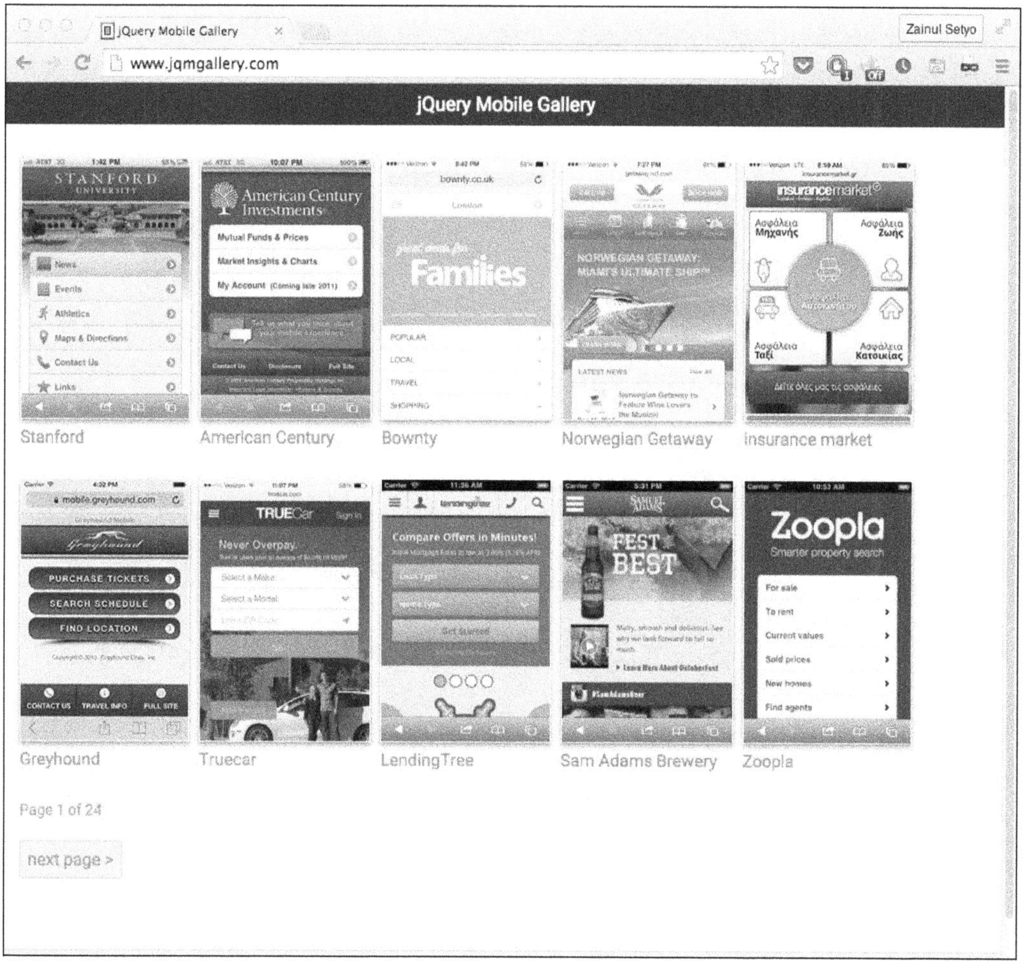

11
Extending PhoneGap with Plugins

In this chapter, we will cover the following recipes:

- Extending your Cordova Android application with a native plugin
- Extending your Cordova iOS application with a native plugin
- The plugin repository

Introduction

Along with providing developers with an incredibly easy yet powerful way to build native mobile applications using HTML, CSS, and JavaScript, the PhoneGap framework also gives us the ability to further extend the functionality by creating native plugins that can interact in more detail with device features and functions that are not already exposed through the existing Cordova plugin API.

By creating native plugins, we can enhance the already vast list of methods available through the plugin API, or build totally unique features, all of which will be made available to call and process responses from JavaScript methods.

For some, the thought of writing code that is native to the device platform that they wish to enhance may be a little daunting. If you haven't had any prior exposure to or experience in these languages, they may be perceived as a form of an unwelcome paradigm shift.

Luckily, the Cordova API simplifies this process as much as possible and essentially breaks it down into two core related parts: some JavaScript code for use within your HTML applications, and a corresponding native class for performing actions and processes in the native code. You will also have to edit and amend some XML to inform Cordova about your plugin and grant permissions for its use, which is also incredibly simple. Cordova will manage the communication between the two parts for you, so all that you have to worry about is building your awesome applications.

Plugins compromise a single JavaScript API along with the corresponding native code for each supported platform. So, we will have various pieces of native code, but we can call native code using a single JavaScript interface.

Extending your Cordova Android application with a native plugin

Android devices have been proven to be developer friendly insofar as they easily allow us to test and deploy unsigned applications throughout the development process.

How to do it...

In this recipe, we will create a native Android plugin for our Cordova application, using command-line tools as suggested by Cordova:

1. Create a new directory named `helloplugin`. Then, create two blank directories, www and `src`, from the root. Next, create an empty file called `plugin.xml`. Now, we have a Cordova plugin project to get started, as shown here:

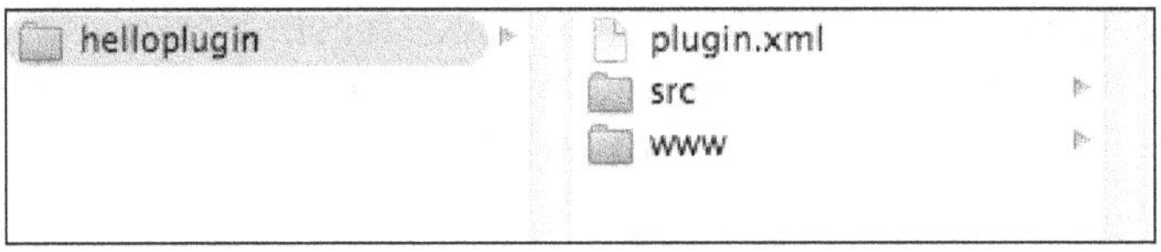

2. We open `plugin.xml` and paste the following code for basic information about the plugin that we are going to make:

```
<?xml version="1.0" encoding="utf-8"?>
<plugin xmlns="http://www.phonegap.com/ns/plugins/1.0"
        id="com.example.hello"
        version="0.7.0">

  <name>Hello</name>
```

```
<engines>
  <engine name="cordova" version=">=3.4.0"/>
</engines>

</plugin>
```

3. Create a new `hello.js` file inside the `www` directory. We have to make a reference from the newly created `hello.js` file to `plugin.xml`:

```
<engines>
    <engine name="cordova" version=">=3.4.0"/>
</engines>

<asset src="www/hello.js" target="js/hello.js"/>

<js-module src="www/hello.js" name="hello">
    <clobbers target="hello" />
</js-module>
```

4. Then, we need to create native Android code. Inside the `src` directory, create another directory named `android`. Inside the `android` directory, create an empty file named `Hello.java`. Again, we have to make a reference to the newly created file in `plugin.xml`:

```
  <js-module src="www/hello.js" name="hello">
    <clobbers target="hello" />
  </js-module>

  <platform name="android">

    <config-file target="res/xml/config.xml" parent="/*">
      <feature name="Hello">
        <param name="android-package"
value="com.example.plugin.Hello"/>
      </feature>
    </config-file>

    <source-file src="src/android/Hello.java" target-dir="src/com/
example/plugin/"/>
  </platform>

</plugin>
```

5. Now, we have completed the plugin information through the `plugin.xml` file. Open `src/android/Hello.java` and create a new class, called `Hello`, that extends from `CordovaPlugin`:

```
public class Hello extends CordovaPlugin {

}
```

6. Add the package named `com.example.plugin` to the class, and import the basic requirements for `CordovaPlugin`:

```
package com.example.plugin;

import org.apache.cordova.*;
import org.json.JSONArray;
import org.json.JSONException;

public class Hello extends CordovaPlugin {

}
```

7. Then, we will override the `CordovaPlugin` `execute` method to suit our own implementation:

```
public class Hello extends CordovaPlugin {
  @Override
    public boolean execute(String action, JSONArray data, 
CallbackContext callbackContext) throws JSONException {

    }
}
```

8. We will create a `greet` action. This action will add the `Hello` string from the given string input:

```
public boolean execute(String action, JSONArray data, 
CallbackContext callbackContext) throws JSONException {

    if (action.equals("greet")) {

        String name = data.getString(0);
        String message = "Hello, " + name;
        callbackContext.success(message);

        return true;

    } else {
```

```
            return false;

        }
    }
```

9. We have now completed the Java for native Android code. The next step is to create a JavaScript interface. Open `www/hello.js`. We will add function invoke `greet` action defined in our previous Java code:

```
module.exports = {
    greet: function (name, successCallback, errorCallback)
{
        cordova.exec(successCallback, errorCallback,
"Hello", "greet", [name]);
    }
};
```

10. We are done creating an Android plugin, and we are ready to add our `Hello` plugin to a Cordova application. First, we have to create a new Cordova application and add the Android platform to the project:

```
cordova create hello com.example.hello HelloWorld
cd hello
cordova platform add android
```

11. To add the plugin, we can simply use `plugman`. If you haven't installed `plugman` yet, you can install it by running `npm install -g plugman`. Once `plugman` is installed, run the `plugman` command from the terminal:

```
plugman install --platform android --project hello/platforms/
android --plugin helloplugin
```

12. Once the plugin has been installed, head back to the Cordova application and open `www/index.html`. Clean it up and add the JavaScript reference of `hello.js` below the `cordova.js` reference and above the `js/index.js` reference:

```
<!DOCTYPE html>
<html>
    <head>
        <meta charset="utf-8" />
        <meta name="format-detection"
content="telephone=no" />
        <meta name="msapplication-tap-highlight"
content="no" />
        <!-- WARNING: for iOS 7, remove the width=device-width and
height=device-height attributes. See https://issues.apache.org/
jira/browse/CB-4323 -->
```

```
        <meta name="viewport" content="user-scalable=no,
initial-scale=1, maximum-scale=1, minimum-scale=1,
width=device-width, height=device-height, target-
densitydpi=device-dpi" />
        <link rel="stylesheet" type="text/css"
href="css/index.css" />
        <title>Hello World</title>
    </head>
    <body>

        <script type="text/javascript"
src="cordova.js"></script>
        <script type="text/javascript"
src="hello.js"></script>
        <script src="js/index.js"></script>
    </body>
</html>
```

13. The next step is to call the JavaScript interface during the `onDeviceReady` event. Open `js/index.js` and add the following code to `onDeviceReady`:

```
var app = {
    // Application Constructor
    initialize: function() {
        this.bindEvents();
    },
    bindEvents: function() {
        document.addEventListener('deviceready',
this.onDeviceReady, false);
    },
    onDeviceReady: function() {
        app.receivedEvent('deviceready');
        var success = function(message) {
            alert(message);
        }

        var failure = function() {
            alert("Error calling Hello Plugin");
        }

        hello.greet("World", success, failure);
    },
    receivedEvent: function(id) {
        var parentElement = document.getElementById(id);
        var listeningElement =
parentElement.querySelector('.listening');
```

```
        var receivedElement = parentElement.querySelector('.
received');

        listeningElement.setAttribute('style',
'display:none;');
        receivedElement.setAttribute('style',
'display:block;');

        console.log('Received Event: ' + id);
    }
};
app.initialize();
```

14. Build and run the project:

    cordova build android

    cordova run android

    We will get an alert box like this:

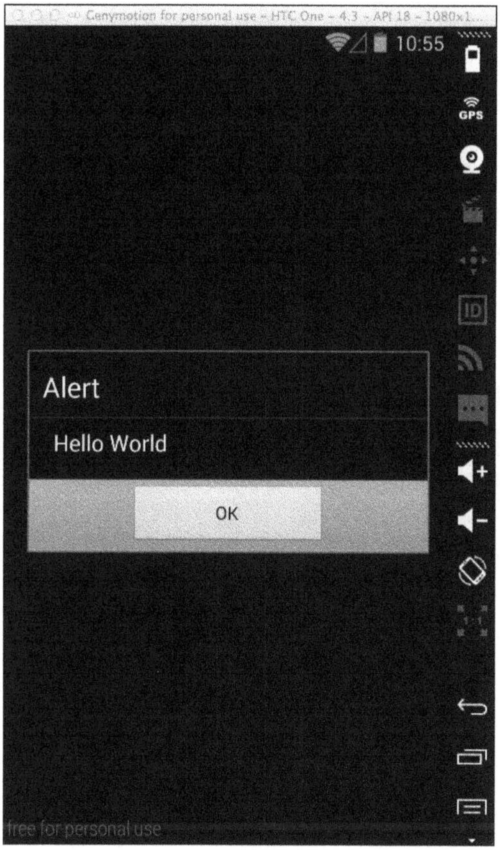

How it works...

We started by creating a new plugin project. There is no generator available, so we ourselves created the `plugin.xml` file and the `www` and `src` directories to hold our project files. We created the plugin definition for general information and the JavaScript module that will be included when the application is built.

Then, we added the platform-specific plugin files. Platform-specific plugin files begin with the `platform` attribute. Inside the `platform` attribute, we declared the package name and filename that contains the native code.

After the plugin directory was prepared, we started writing the Java code required to hold the native code for the Android platform. We created the `Hello` class, which extends `CordovaPlugin`. Then, we implemented the `execute` method to detect the greet action sent from the JavaScript interface. We appended the word `"Hello, "` to the string parameter given by the JavaScript interface.

After we were done with the native Android code, we created the JavaScript interface. Our Cordova application will call and execute the JavaScript code from this interface. The Cordova core enables the JavaScript interface to call native code and then return the call to JavaScript.

With the help of `plugman`, we added the `hello` Cordova plugin to our `hello` application. The `plugman` makes the process of Cordova plugin development easier. We don't need to register our plugin on any online Git repository.

> For more information about developing your own Cordova plugin, visit `https://cordova.apache.org/docs/en/4.0.0/guide_hybrid_plugins_index.md.html`.

Extending your Cordova iOS application with a native plugin

How to do it...

In this recipe, we will create a native iOS plugin for our Cordova application using command-line tools, as suggested by Cordova. We will implement the plugin for Android that we created in the previous recipe to work with iOS:

1. Change the working project to `helloplugin`, open `plugin.xml`, and add information about the new iOS platform:

```
  <platform name="ios">

    <config-file target="config.xml" parent="/widget">
      <feature name="Hello">
        <param name="ios-package" value="HWPHello" />
      </feature>
    </config-file>

    <header-file src="src/ios/HWPHello.h" target-
dir="HelloPlugin"/>
    <source-file src="src/ios/HWPHello.m" target-
dir="HelloPlugin"/>
  </platform>

</plugin>
```

2. Then create two blank files, `HWPHello.h` and `HWPHello.m`, inside the `src/ios` directory. Both files will contain our native Objective-C code for the iOS platform.
3. Open `HWPHello.h` and paste the following code in it:

```
#import <Cordova/CDV.h>

@interface HWPHello : CDVPlugin

- (void) greet:(CDVInvokedUrlCommand*)command;

@end
```

4. We have created the `HWPHello` interface, so we need to create the class implementation. Open `HWPHello.m` and paste this code in it:

```
#import "HWPHello.h"

@implementation HWPHello

- (void)greet:(CDVInvokedUrlCommand*)command
{

    NSString* callbackId = [command callbackId];
    NSString* name = [[command arguments] objectAtIndex:0];
    NSString* msg = [NSString stringWithFormat: @"Hello,
%@", name];
```

```
    CDVPluginResult* result = [CDVPluginResult
    resultWithStatus:CDVCommandStatus_OK
                                  messageAsString:msg];

    [self success:result callbackId:callbackId];
}

@end
```

5. We have created our `Hello` Cordova plugin for the iOS platform. The next step is to change the directory to the `hello` Cordova app and then add the iOS platform to the project:

   cd hello

   cordova platform add ios

6. Then, we use `plugman` to install the iOS plugin that we created in our `hello` Cordova project:

   plugman install --platform ios --project hello/platforms/ios --plugin helloplugin

7. Build and run the project:

   cordova build ios

   cordova emulate ios

8. We will have our hello Cordova application running and displaying an alert box, as follows:

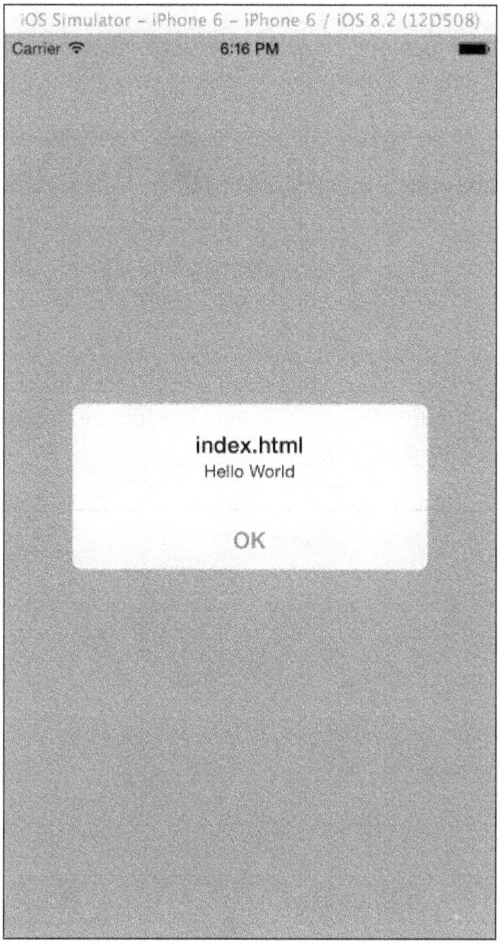

How it works...

In this recipe, we continued to work with the `hello` Cordova plugin. We created the plugin implementation for iOS. We began by adding a platform element to `plugin.xml`. Inside the `platform` element, we added references to native Objective-C code. The code consists of the `HWPHello.h` header and the `HWPHello.m` implementation.

We included the dependencies of `Cordova/CDV.h` inside the `header` file. Then we created the `HWPHello` interface, which extends from `CDVPlugin`. Inside the `HWPHello` interface, we created a new `void` method named `greet`. The `greet` method accepts `CDVInvokedUrlCommand` as a parameter. After we were done creating the header file, we created the implementation of the interface in `HWPHello.m`. Inside the `greet` method, we appended the word `"Hello, "` to the string inputted by the JavaScript interface.

Since we have created a JavaScript implementation in the previous recipe, we don't need to create a new JavaScript implementation. All we have to do is add the plugin to our hello Cordova project with the help of `plugman`.

The plugin repository

Cordova enables developers to create their own plugin to extend hybrid application capabilities. This plugin system enables hybrid applications to get native capabilities. Almost any native functionality can be implemented using plugins.

Sometimes, creating our own plugin will take more time. We have to know each of the platform's languages, create code specific to each platform, and then create a JavaScript interface to make it work with an HTML application. It would be easier for us to use an existing plugin or an open source Cordova plugin.

Luckily, there is a centralized repository of Cordova plugins. We can search for the plugin we need and then add it to our Cordova application.

How to do it...

To search and use a Cordova plugin from the repository, follow these steps:

1. Open `http://plugins.cordova.io/#/`, and you will see a search box to start searching for a plugin, like this:

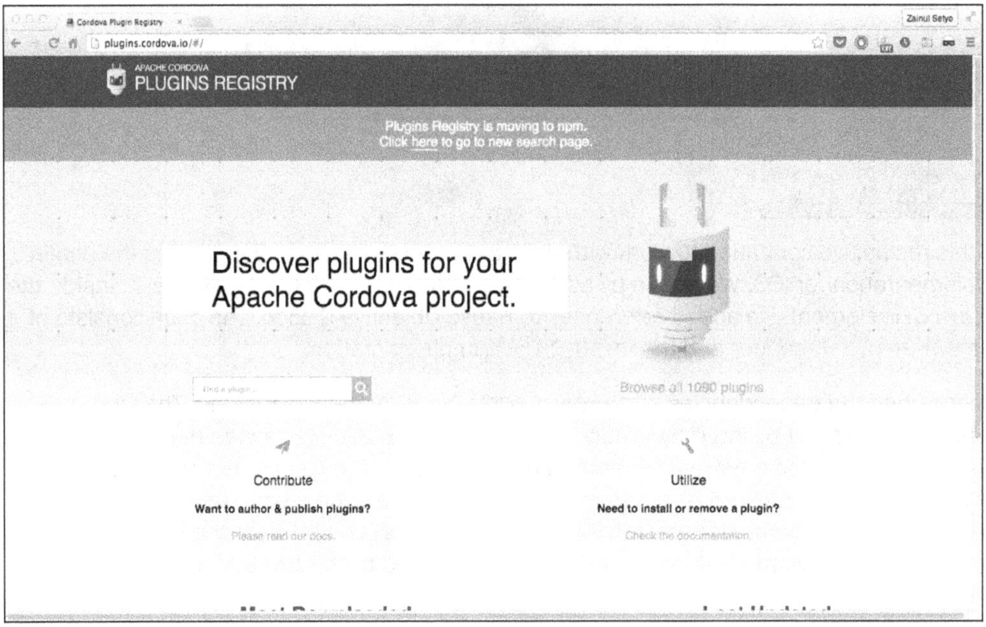

2. Let's find a plugin for implementing local notifications for the Cordova application. We type the `notification` keyword in the search box, and then hit *Enter* so that we can see a list of notification plugins available, along with a brief description of each plugin, as shown here:

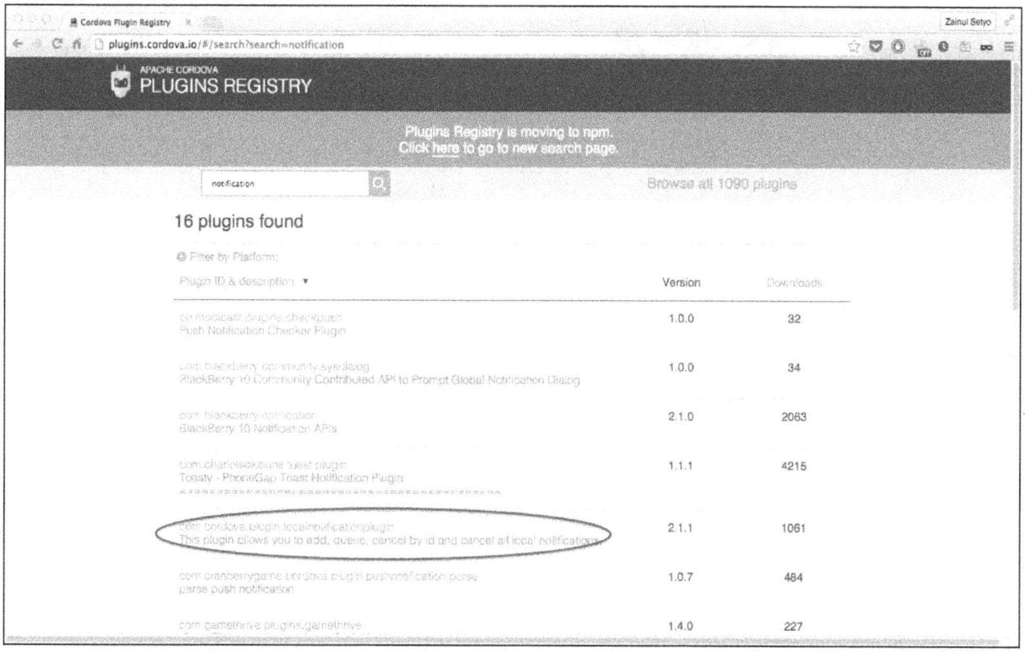

3. Now we find the local notification plugin named with the package name as com.cordova.plugin.localnotificationplugin. Open the link and you will see the plugin information page, as shown in the following screenshot:

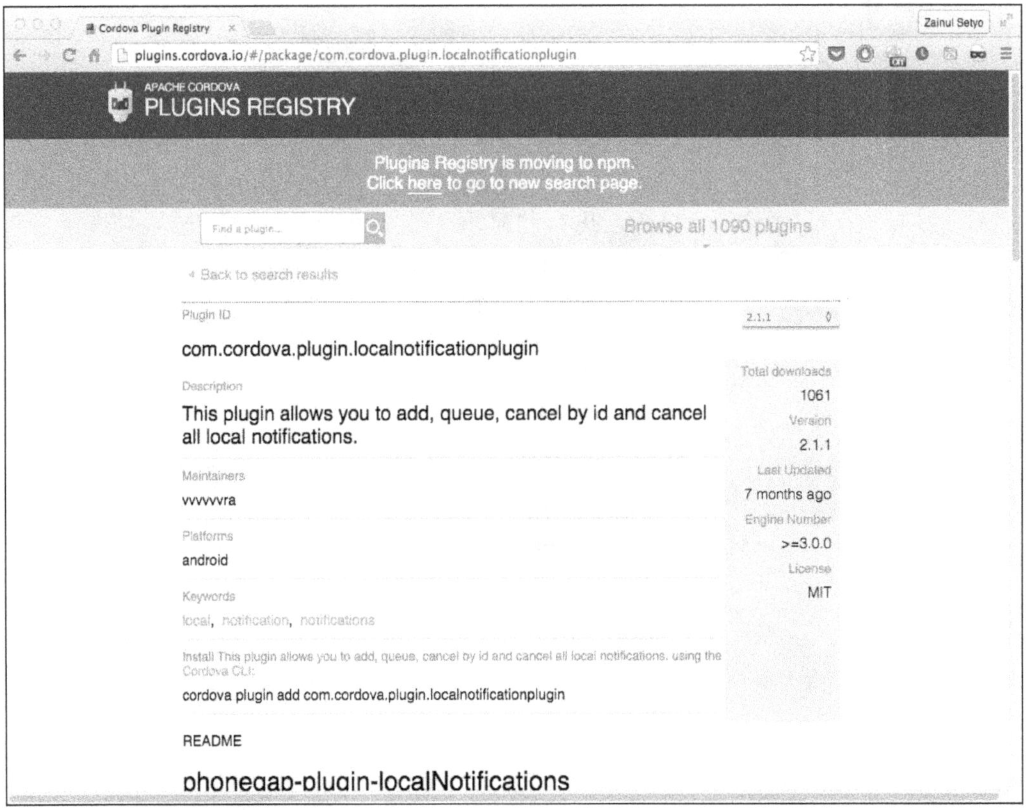

4. All we have to do is add the plugin to our application:

```
cordova plugin add com.cordova.plugin.localnotificationplugin
```

5. After we are done adding the plugin to the application, we only need to call the JavaScript API, as the plugin description says.

How it works...

We started by visiting the official Cordova plugin registry at http://plugins.cordova.io. Then we searched for the local notification plugin using the notification keyword. Once we found the plugin, we opened the plugin page. Then we added the plugin to the Cordova application using the cordova plugin add <package name> command.

12
Development Tools and Testing

In this chapter, we will cover the following recipes:

- Downloading Cordova
- Using the command line to create a new iOS Cordova application
- Debugging the iOS Cordova application using Safari Web Inspector
- Using Android Studio to develop Android Cordova applications
- Using Adobe Dreamweaver to develop Cordova applications
- Using the PhoneGap Build service

Introduction

To successfully create your Cordova applications, it is really important to set up the correct development environment—one that suits the requirements of your application, your personal development style, and the tools and features that you may need to use, and one that is compatible with your local development machine's operating system.

In this chapter, we will investigate some of the options available for setting up your local environment with development tools in order to assist you and help make developing your mobile applications even easier.

Downloading Cordova

Before we develop any Cordova applications, we need to download a copy of the framework, of course.

How to do it...

In this recipe, we will download the Cordova framework to make sure that we have the framework available to start local development:

1. Cordova needs Node.js to be installed first. Head over to `https://nodejs.org/en/download/` and download the installer for your machine, as shown here:

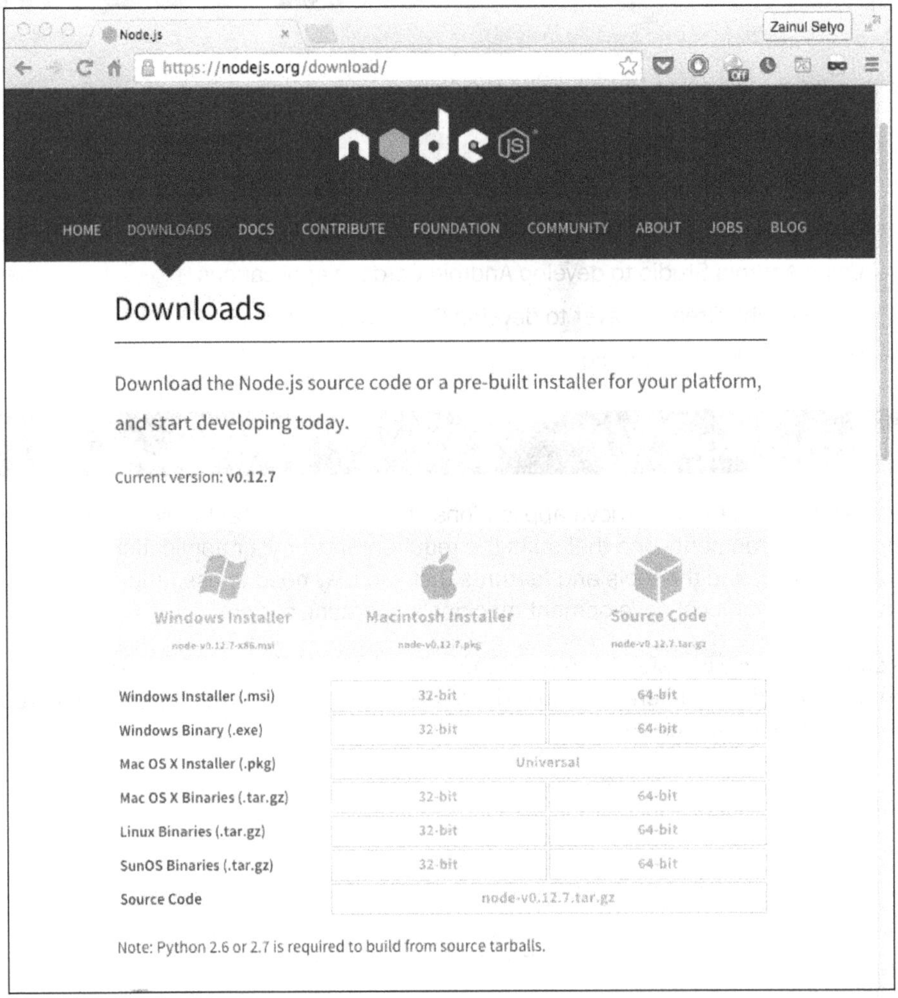

2. Download Cordova by running the following command on your terminal or Command Prompt:

   ```
   npm install -g cordova
   ```

3. Check your Cordova installation by running `cordova -v`. You will see something like this if Cordova is installed correctly:

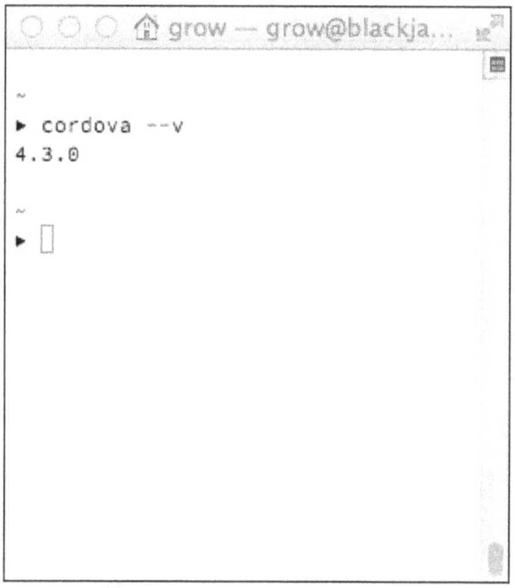

How it works...

We installed Node.js from the official site. Then we used the Node Package Manager to download and install Cordova.

Using the command line to create a new iOS Cordova project

A streamlined workflow is something that can benefit us all greatly by speeding up our processes and reducing the amount of manual work needed to complete a task.

How to do it...

In this recipe, we will explore the command-line tools available in Cordova for creating and running iOS applications from the terminal application:

1. Create a new Cordova iOS project by running the following command:

   ```
   cordova create hello com.example.hello HelloWorld
   ```

2. Change the directory to the newly created project:

   ```
   cd hello
   ```

3. Add the iOS platform to the project:

   ```
   cordova platform add ios
   ```

4. Open the project directory in the explorer, and you should see something like this:

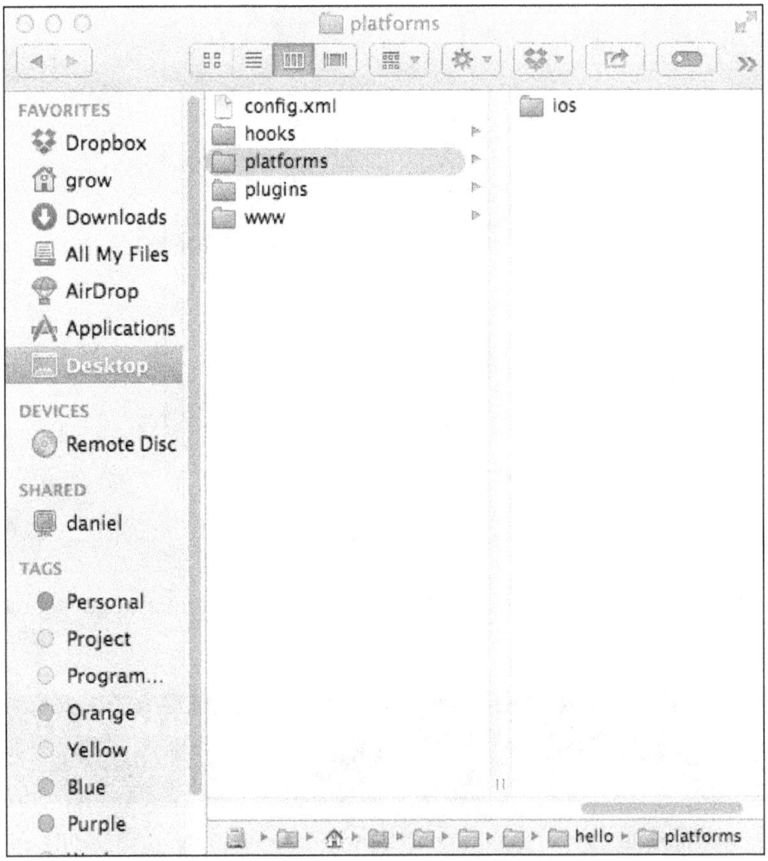

5. Open the `/platform/ios` directory and double-click on `HelloWorld.xcodeproj`. So, the project will be opened in Xcode, as shown here:

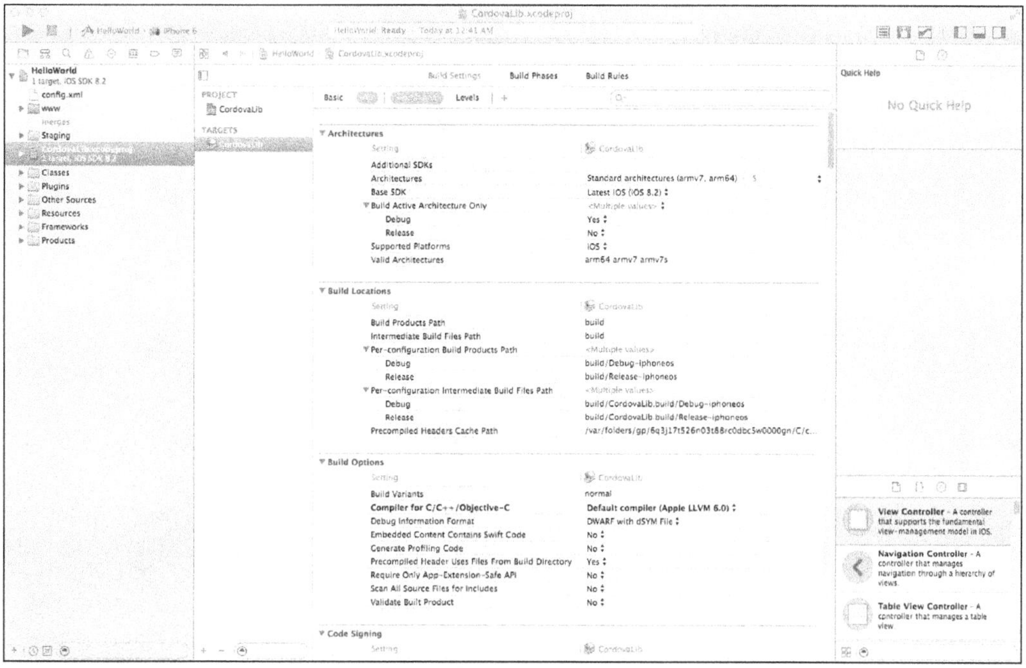

6. If we choose to run the newly opened project via Xcode on a simulated device, the output will look something like this:

How it works...

The command-line tools available in Cordova greatly simplify the tasks involved in creating a project. With a simple command, we can generate a complete project—complete with the `www` directory linked at a project level, and ready to open up in Xcode to continue our development.

The `create` command accepts three parameters:

- The path to your iOS project
- The name of the package
- The project name

There's more...

The usage of the command line is incredibly powerful, and it can allow developers to automate the creation of a project by running a very simple script. However, that's not the only tool available for use through the command line.

> To see all the available commands in command-line tools, visit `http://cordova.apache.org/docs/en/5.1.1/`.

Running the application on the iOS Simulator

We can run the `emulate` script from the command line via the Terminal app to launch our application on a simulated device.

For the emulation to work successfully on OS X, you will need to install a command-line utility tool called **ios-sim**, an open source project that launches the built application on the iOS simulator. You can download ios-sim from the GitHub repository at `https://github.com/phonegap/ios-sim`.

The `readme` file in this repository has short, detailed instructions explaining how to install ios-sim on your machine.

Once the installation is done, simply run the `emulate` script to load the application onto the iOS Simulator. To do so, simply type the path to the `cordova` directory within your project folder and run the `cordova emulate ios` script.

When you run this command for the first time, the script will ascertain whether or not you currently have a successfully built version of the application. If not, it will ask you whether or not you would like the build process to take place.

Debugging your application

When running an application on a simulator using Xcode, we are able to catch the `console.log` message directly from Xcode. Another option is to use Safari to debug and inspect your running application.

Debugging the iOS Cordova application using Safari Web Inspector

Debugging iOS Cordova can be done just like debugging a normal web application. We can catch network activities, inspect elements, and see the console logging.

How to do it...

To start debugging a iOS Cordova application, follow these steps:

1. Open and run the `HelloWorld` project created before using Xcode.
2. While the application is running on the simulator, open Safari. Go to **Develop** | **iOS Simulator** | **index.html**, as shown in the following screenshot:

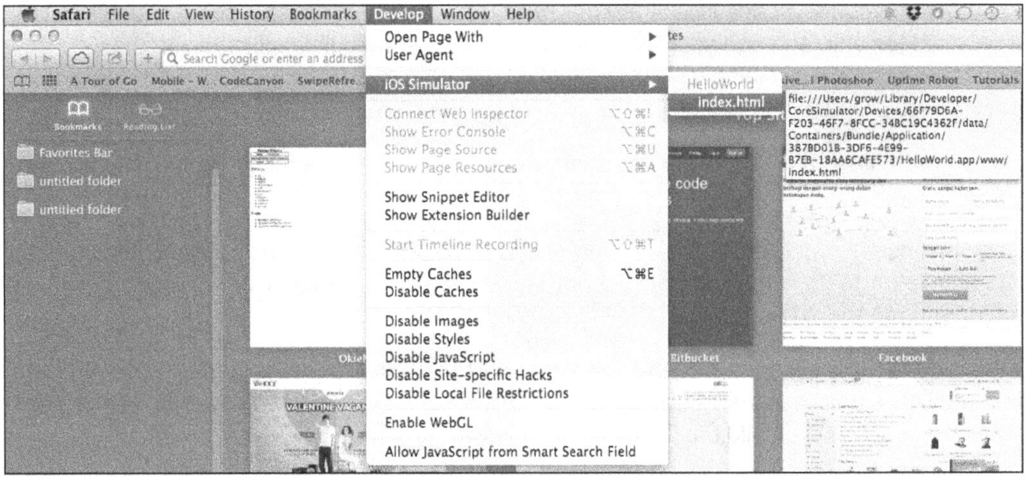

3. We can inspect the element and check out **Local Storage** and **Session Storage** by choosing the **Resources** tab, as follows:

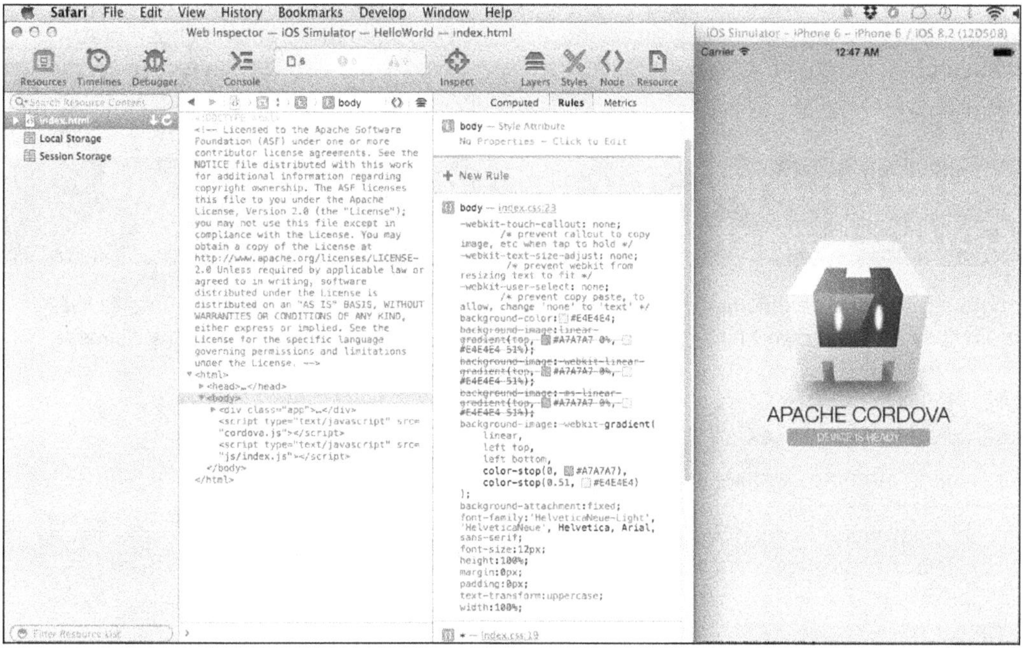

4. We can see the network activities on the **Timelines** tab, as shown in this screenshot:

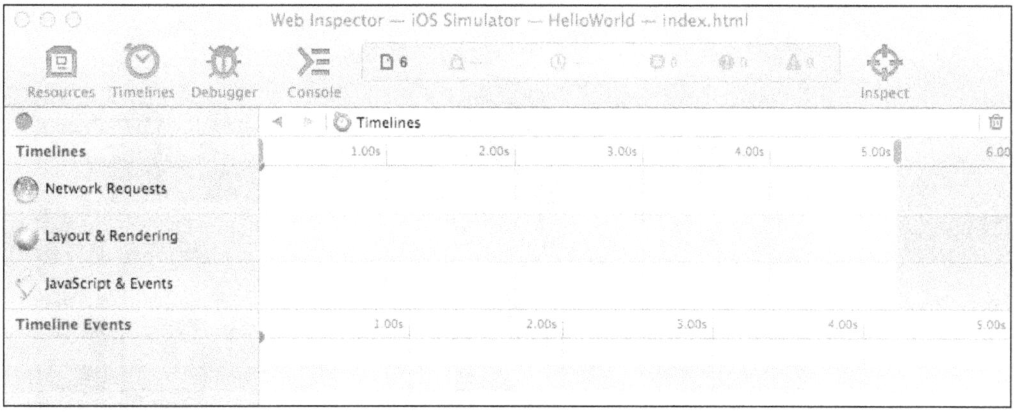

5. To see the output on the browser's JavaScript console, click on the **Console** tab.

How it works...

We opened the iOS Cordova application in Xcode. Then we ran the application on a simulator. While the application was being run on the simulator, we opened Safari and chose **Web Inspector**. Safari connects to the emulator's Safari to make the debugging process easier. The debugging process is practically the same as debugging common web or HTML5 applications.

Using Android Studio to develop Android Cordova applications

Android Studio is the official IDE for developing Android applications, based on IntelliJ IDEA. It provides an easier workflow to develop Android applications compared to using Eclipse.

Getting ready

Opening an Android Cordova application is very simple:

1. Download Android Studio from `http://developer.android.com/sdk/index.html`, as shown in this screenshot:

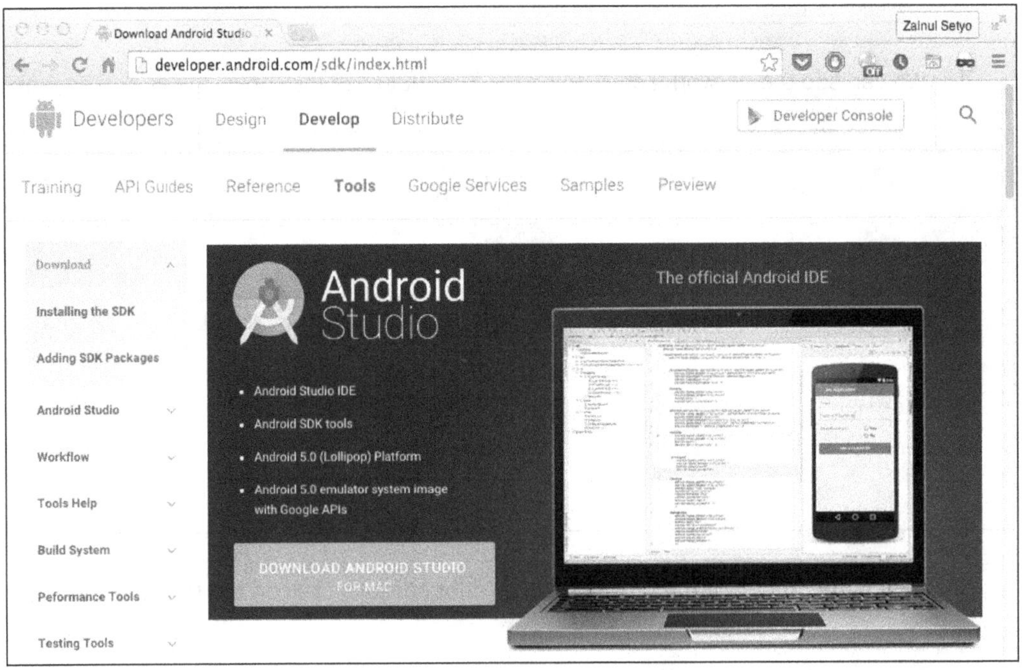

2. Launch Android Studio.
3. Select **Import project (Eclipse ADT, Gradle, etc.)**, as shown in the following screenshot:

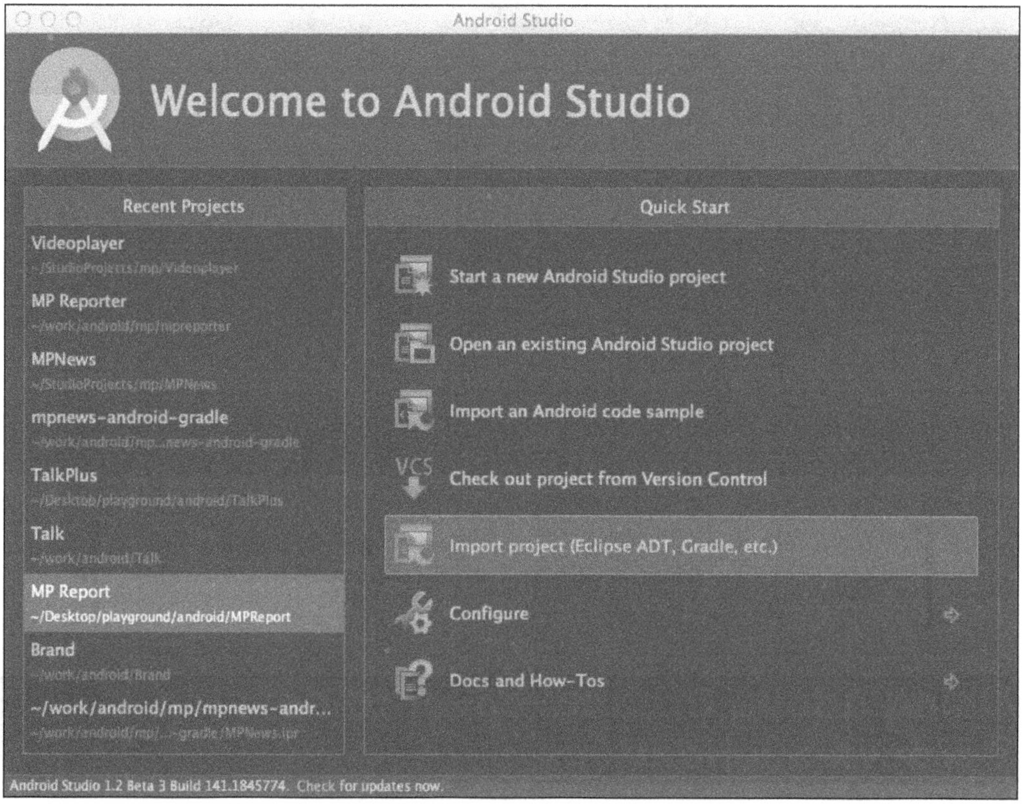

4. For the Gradle Sync question, you can simply answer **Yes**.
5. Select the location where the `android` platform is stored (`phonegap-cookbook/chapter12/hello/platforms/android`).

6. Your application is ready and can be run directly from Android Studio.

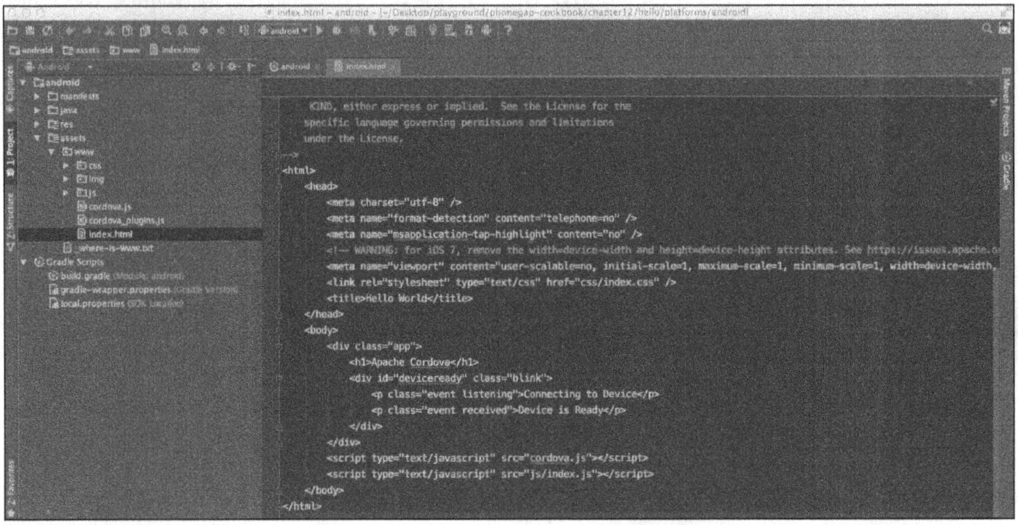

> For more information about building using Android Studio, visit `http://developer.android.com/tools/building/building-studio.html`.

Using Adobe Dreamweaver to develop Cordova applications

Adobe Dreamweaver has long been a favorite tool of web designers and developers. Dreamweaver CS5 included features that assisted building mobile applications and the ability to simulate and package PhoneGap applications without having to worry about the underlying framework library.

Dreamweaver CS6 took this one step further and integrated an automated Build service in the form of PhoneGap Build.

Getting ready

To use the PhoneGap Build service, you will first need an active PhoneGap Build service account, which is free and incredibly easy to set up:

1. Head over to `https://build.phonegap.com/people/sign_in` to begin the registration process. You can sign in using either your Adobe ID or your GitHub account details, as shown here:

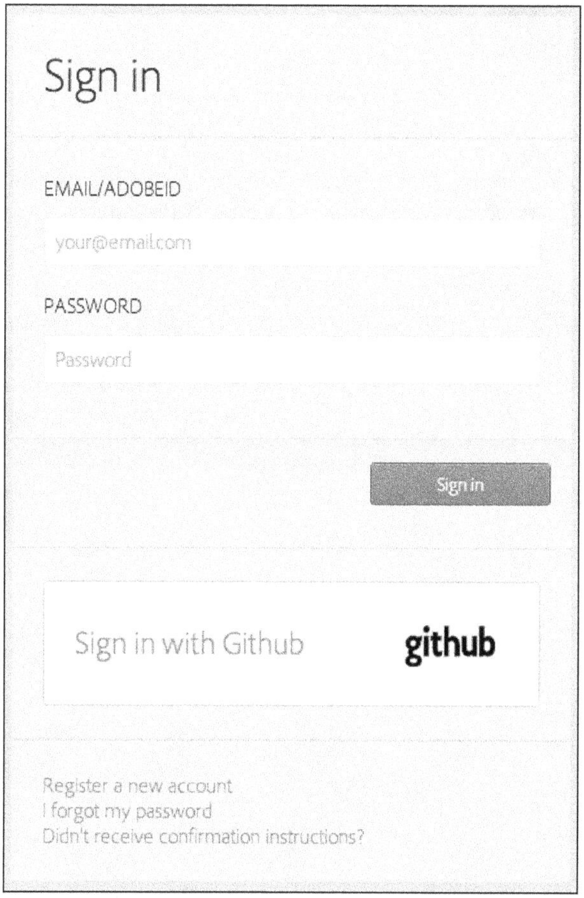

That's it! It's time to start building an application using Dreamweaver CS6.

> If you do not have a copy of Dreamweaver CS6, you can download a full-featured trial version from `http://www.adobe.com/uk/products/dreamweaver.html`.

How to do it...

In this recipe, we will use Dreamweaver CS6 to create a simple application, which we will pass to the PhoneGap Build service to build for us:

1. To manage our Cordova application in Dreamweaver CS6, we need to create a new Dreamweaver site definition, pointing the project directory to our preferred location on our development machine.
2. We'll simplify the creation of our application's structure and download a project template that contains all the core files that we will need. Head over to `https://github.com/phonegap/phonegap-start` and click on the **ZIP** button to download a compressed version of the code. Alternatively, you can clone the GitHub project directly on your local machine using the command line.

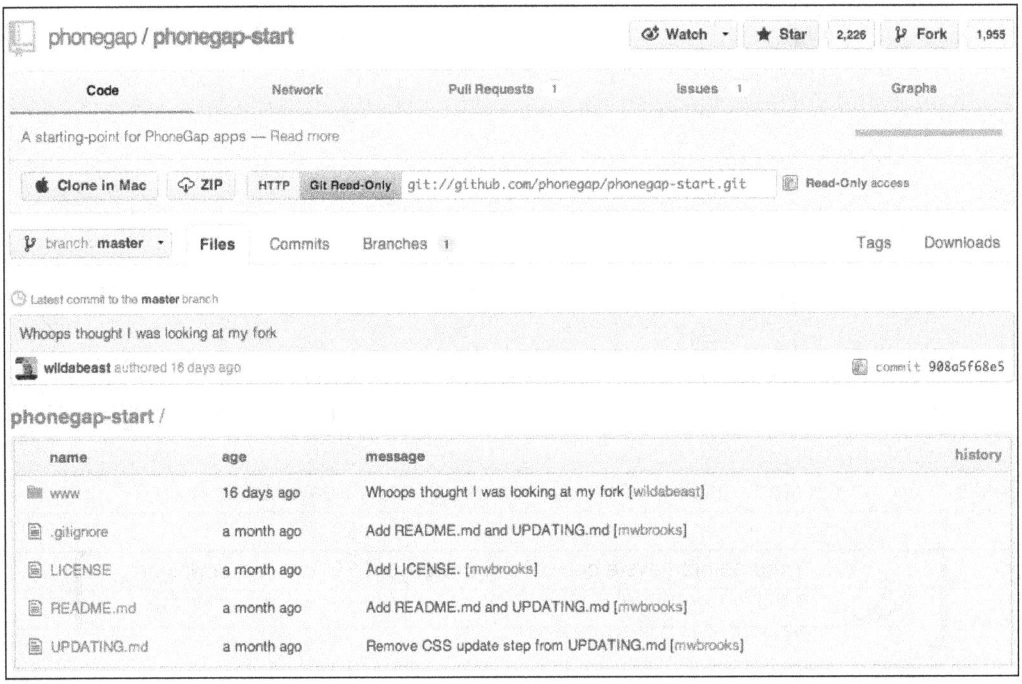

3. Move or copy the extracted files to the location of your Dreamweaver project. The resulting project structure should look something like this:

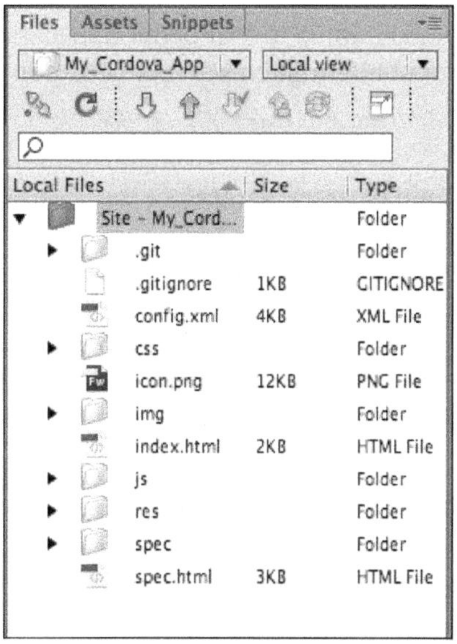

4. Open `index.html` in the editor window. There are some JavaScript assets included in the `head` of the document, one of which is called `cordova-2.0.0.js`. However, you may notice that the file does not actually exist in the project itself.

```
<html>
    <head>
        <meta http-equiv="Content-Type" content="text/html; charset=UTF-8" />
        <meta name="format-detection" content="telephone=no" />
        <meta name="viewport" content="user-scalable=no, initial-scale=1, maximum-scale=1, minimum-scale=1, width=device-width,
height=device-height, target-densitydpi=device-dpi" />
        <link rel="stylesheet" type="text/css" href="css/index.css" />
        <title>Hello World</title>
    </head>
    <body>
        <div class="app">
            <h1>PhoneGap</h1>
            <div id="deviceready">
                <p class="status pending blink">Connecting to Device</p>
                <p class="status complete blink hide">Device is Ready</p>
            </div>
        </div>
        <script type="text/javascript" src="cordova-2.0.0.js"></script>
        <script type="text/javascript" src="js/index.js"></script>
        <script type="text/javascript">
            app.initialize();
        </script>
    </body>
</html>
```

> When dealing with the PhoneGap Build service, you do not need to have a local copy of the PhoneGap/Cordova JavaScript file, but you do need to reference it in the document. This is because the service needs it to successfully build your application.

5. Open the `config.xml` file in the editor window, which holds the configuration details for our application. Change the `id` attribute value to a reverse-domain value specific to you.
6. We'll also change the `name` value to that of our application, and change the `author` node to set the developer information with our URL, e-mail address, and name:

```
<?xml version="1.0" encoding="UTF-8"?>
<widget
xmlns=http://www.w3.org/ns/widgets
xmlns:gap=http://phonegap.com/ns/1.0
id="com.coldfumonkeh.mycordovaapp"
version="1.0.0">

<name>My Cordova Application</name>
<author href="http://www.monkeh.me" email="me@monkeh.me">Matt
Gifford</author>
```

7. We can request that our application be built using a particular version of the PhoneGap library. Although the Build service will use the default version (currently 2.0.0), you can also enforce this by setting a `preference` tag to confirm that you want to use that version:

```
<preference name="phonegap-version" value="2.0.0" />
```

8. Then, there is the `feature` node in the `config.xml` file, which specifies the features you want your application to use and have access to. This sample application requests access to the device API. Let's add two more `feature` nodes to request access to the `geolocation` and `network` connectivity APIs, like this:

```
<feature name="http://api.phonegap.com/1.0/geolocation"/>
<feature name="http://api.phonegap.com/1.0/network"/>
```

> For detailed information about all the parameters and values available for use in the `config.xml` file, visit the PhoneGap Build documentation at `https://build.phonegap.com/docs/config-xml`.

9. Save the `config.xml` file, and open the PhoneGap Build service window by going to **Site** | **PhoneGap Build Service** | **PhoneGap Build Service** from the main application menu. This will open the Build service window in Dreamweaver. If this is the first time you are running this process within Dreamweaver, you will be asked to enter your PhoneGap Build account credentials to log in and use the remote service, as shown in the following screenshot:

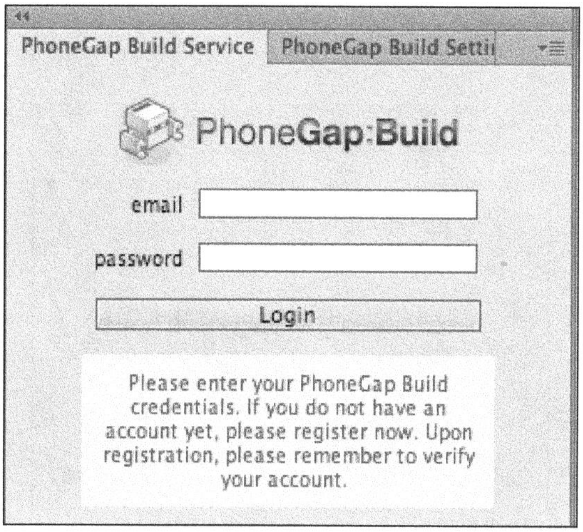

10. Once logged in, you will be presented with the Build service panel, which will inform you that a project settings file needs to be built for this application. Ensure that **Create as a new project** is selected in the drop-down box, and click on **Continue** to start the build process, as shown in this screenshot:

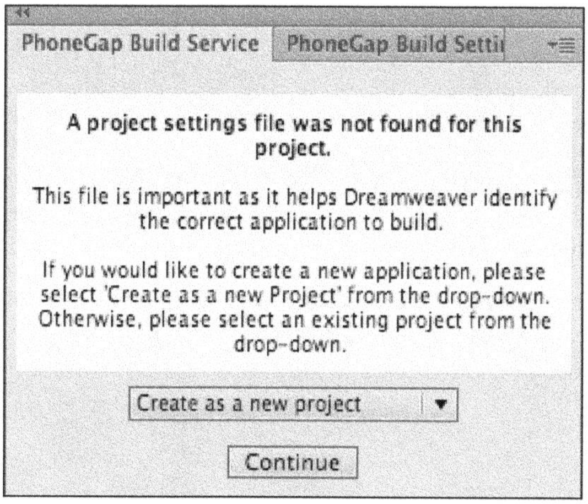

11. Dreamweaver will now submit your application to the remote PhoneGap Build service. You can view the process of each build from the panel in the editor.

> If you receive any build errors, make sure that the `id` value in the `config.xml` file does not contain any underscores or invalid characters, and that there are no spaces within the `xml` node names and values.

12. Once the build process is complete, you will be shown the results for each mobile platform, like this:

13. You will be presented with the option to download the packaged application in the respective native format by clicking on the downward-facing arrow. You can then install the application on your device for testing.
14. If your mobile device has a **Quick Response** (**QR**) Code Reader, you can select the barcode button and scan the resulting image shown to you. This will download the application directly on the device without any need of USB connectivity or transfers.

15. In this example, the iOS application was not built as no signing key was provided in the `config.xml` file.

> For more details on iOS code signing, check out the official documentation from the iOS Developer Library at `http://developer.apple.com/library/ios/#technotes/tn2250/_index.html`.

16. You may also notice that the Android platform have an extra button—the right-facing arrow. Clicking on this button will load the application onto a local emulator for testing purposes.
17. To use emulators on your local machine, Dreamweaver needs to know the location of the SDK libraries. To set these paths, open the **PhoneGap Build Settings** panel, which you can reach by going to **Site** | **PhoneGap Build Service** | **PhoneGap Build Settings** from the main menu, as shown in this screenshot:

18. Include the path to the SDKs that you have installed and wish to use, and click on **Save** to store them.

How it works...

The integration of the PhoneGap Build automated service directly into Dreamweaver CS6 helps manage the packaging and building of your application across a number of device platforms. The build process takes care of the device permissions and inclusion of the correct Cordova JavaScript file for each platform.

This means that you can spend more time developing your feature-rich native application and perfecting your layouts and visuals, without having to handle the various build processes and differences between each platform.

Using the PhoneGap Build service

A benefit of the Cordova project is the ability to create a native application from an HTML file and a little JavaScript, at its very minimum. Adobe Dreamweaver CS6 has the built-in capabilities to interact with and upload your mobile project directly on the remote PhoneGap Build service on your behalf. You can, however, build your application using the service directly via the Web.

Getting ready

To use the PhoneGap Build service, you will first need an active PhoneGap Build service account, which is free and incredibly easy to set up:

1. Head over to `https://build.phonegap.com/people/sign_up` to begin the registration process. You can sign in using either your Adobe ID or your GitHub account details, as shown here:

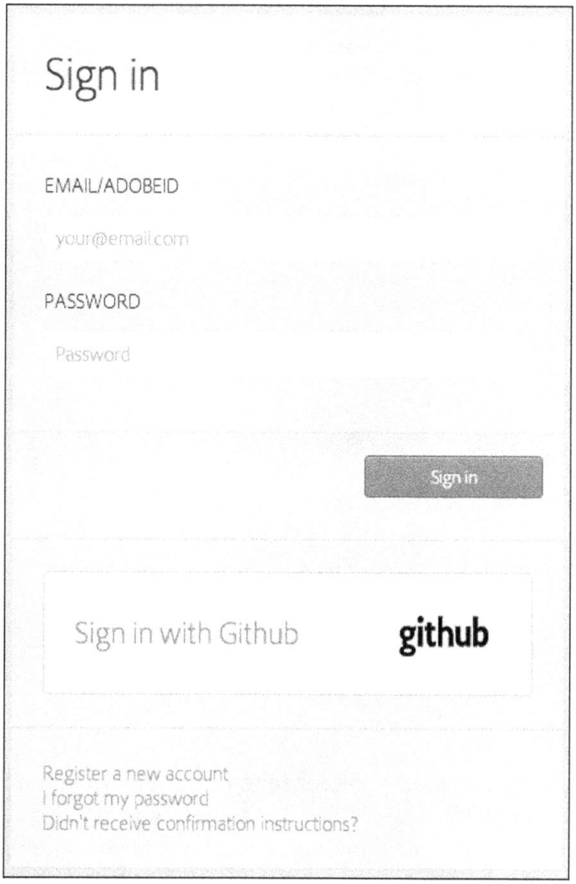

How to do it...

In this recipe we will create a very simple single-file application and upload it to the Build service using the web interface:

1. Create a new `index.html` file, and include the following code in it:

```
<!DOCTYPE html>
<html>
<head>
  <meta name="viewport" content="user-scalable=no,
    initial-scale=1, maximum-scale=1,
    minimum-scale=1, width=device-width;" />
  <title>Cordova Application</title>
```

```
  <script type="text/javascript"
      src="cordova.js"></script>

  <script type="text/javascript">
    function onLoad() {
      document.addEventListener("deviceready",
onDeviceReady, false);
    }

    function onDeviceReady() {
      alert('Cordova has loaded');
    }
  </script>

</head>
<body onload="onLoad()">

  <h1>Welcome to Cordova</h1>

  <p>What will you create?</p>

</body>
</html>
```

In this file, we have included a reference to `cordova.js` in the head of the document. This file does not exist locally, but the Build service requires this reference to successfully build the application. We have created a very simple function that will generate an alert notification window once the device is ready.

2. We create a new `.zip` file that contains the `index.html` file we just created. We will use this archive file to send to the PhoneGap Build service to create our application.
3. Head over to `https://build.phonegap.com/`, and sign in with your account details if you haven't done so already. If this is your first visit, or you have not yet submitted any application to the service, you will be greeted with a form to help you generate your first build.
4. You have the option to create either an open source or a private application. Both options give you the ability to provide code as either a link to a Git repository or an archived `.ZIP` file. For the purposes of this recipe, we will select the **Upload a .zip file** button. We navigate to the archived file that we created earlier, select it, and click on **Open** to proceed. The PhoneGap Build service will instantly upload our code and create a new project, as shown here:

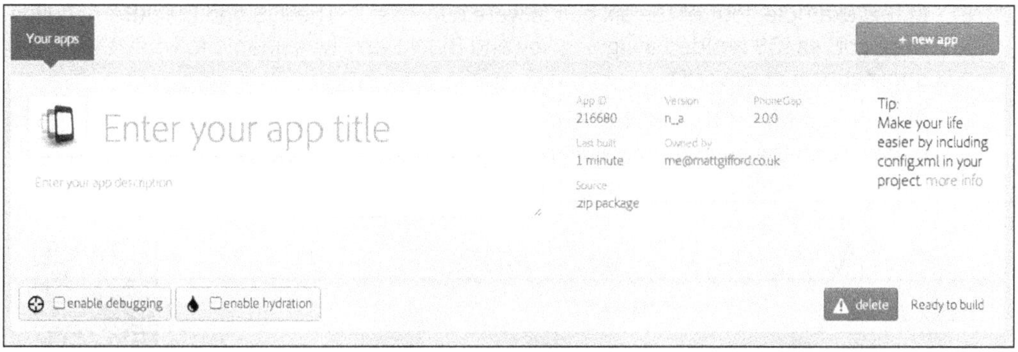

5. As we have not included a `config.xml` file in the upload, the application title and description are placeholders. Click on the **Enter your app title** text and enter the title or name of your application in the input box, making it easily identifiable so that you can distinguish it from any other apps that you may upload. You can also add a description for your application in the same manner.
6. Let's also select the ability to debug our application, as shown in this screenshot:

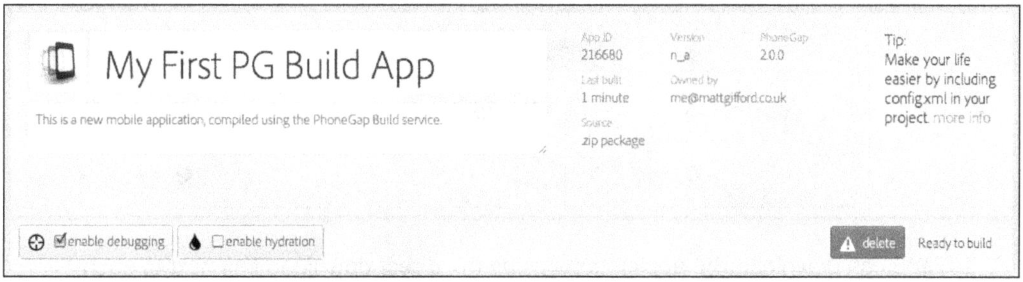

7. Click on the **Ready to build** button in the bottom-right corner of the screen layout. The build process will now begin, and you will be presented with a visual reference to the available platforms and the status of each build, like this:

8. In this example, four of the six builds were successful. The iOS and BlackBerry builds were not, as iOS needed a signing key and BlackBerry was unable to verify the application's passwords.
9. Click on the title of the application, which will take you to a new page with the options to download the successfully built packages, as shown in the following screenshot:

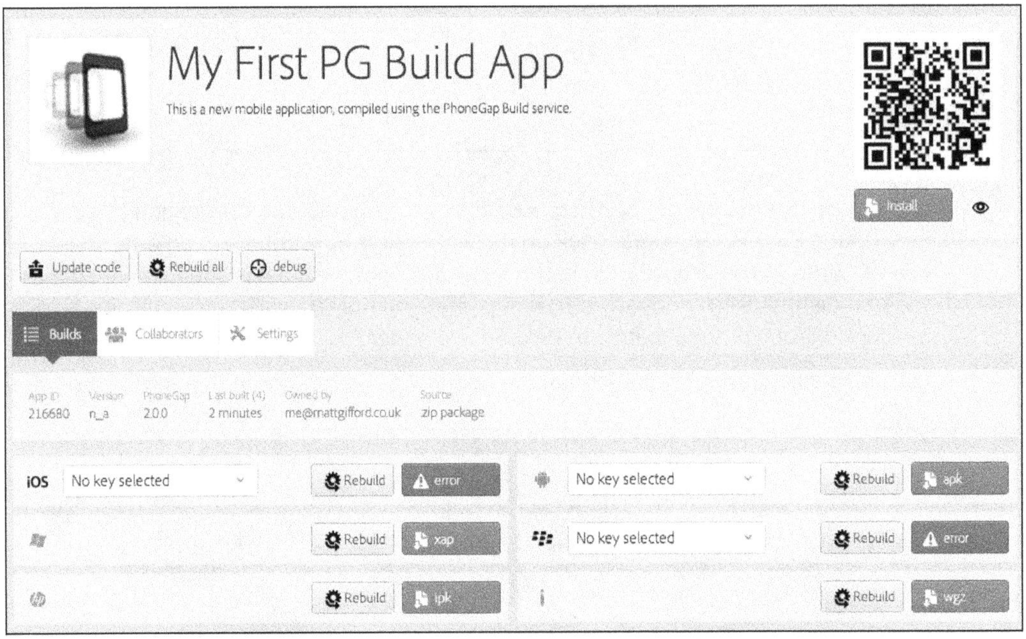

10. You can download the applications using the direct download link next to each successful build listing; or if your device has a QR Code Reader, you can scan the barcode to download and install the application directly on the device.

How it works...

The PhoneGap Build service automates the packaging process for your mobile application across the available device platforms, which greatly simplifies your workflow and the task of preparing configuration files for each individual platform.

There's more...

When we created this sample application using the online Build service, we chose to enable debugging. This is another online service offered by the PhoneGap team, and it makes use of an open source debugging application called **weinre**, which stands for **Web Inspector Remote**.

To see it in action, you need to have the application packaged by the PhoneGap Build service running on a physical device attached to your local development machine, or on a device emulator. Once you have the application running, click on the **debug** button at the top of your project page, as shown here:

This will open a new window or tab in your browser using the `debug.phonegap.com` subdomain, and you should see your connected device in the list, as shown in the following screenshot:

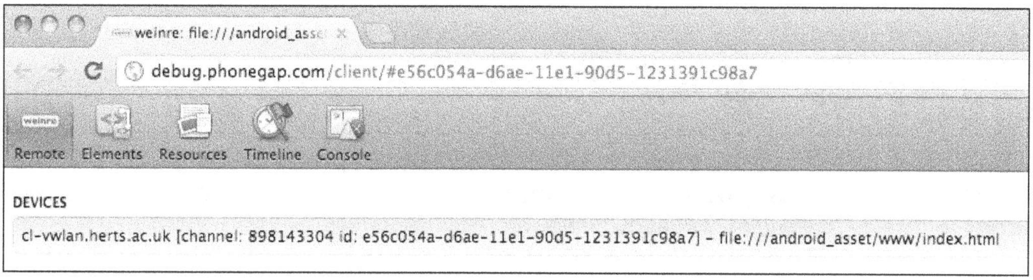

For developers who have used development tools such as Firebug, the tools on offer with weinre should look very familiar. It provides similar functionality to test and debug HTML applications inline, but is designed to work remotely and is exceedingly good at testing apps and pages on mobile devices.

> To find out more about the features and functions available in weinre, check out the official documentation at `http://people.apache.org/~pmuellr/weinre/`.

The PhoneGap Build documentation also provides an introduction to the service with an emphasis on using it with the online Build service, at `https://build.phonegap.com/docs/phonegap-debug`.

Hydrating your application

PhoneGap Build also provides a tool to enhance the workflow and deployment of compiled applications on your devices for testing, called Hydration. This reduces the time taken to compile an application and automatically pushes a new build of a "Hydrated" app onto the device. Hydration can be enabled while creating a new project, or the settings can be updated and applied to an existing project.

To apply Hydration to your current application, select the **Settings** tab within the project view and click on the **enable hydration** button. Then click on the **Save** button to retain the changes. At this point, the application will automatically rebuild as a hydrated app.

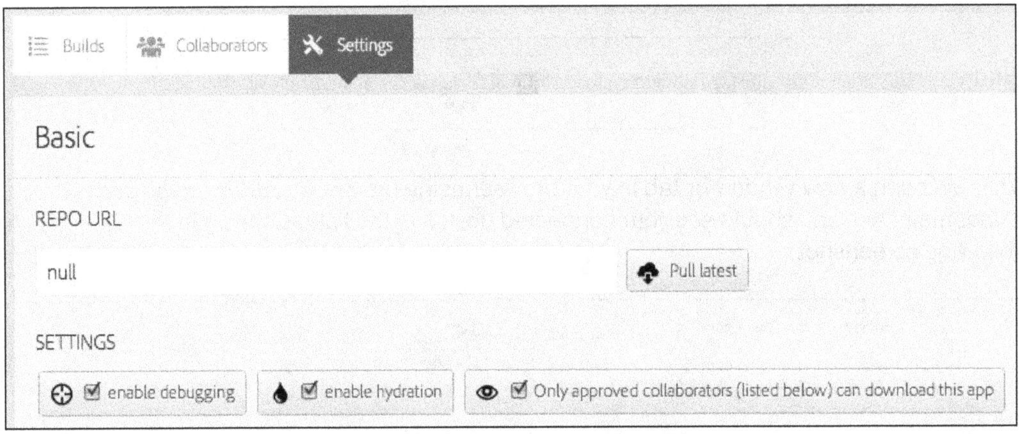

Install or deploy the freshly compiled application on your device, replacing any previous non-hydrated version that you may already be having.

The benefit of such an application is the ease and simplicity of deploying further updates. Once any future code has been updated, deployed, and compiled in the Build service, you can easily update your installed version by restarting the application on the device. The Hydrated application will check on every startup whether an updated version of the app exists, and if so, it will prompt the user with a dialog box offering a chance to update the application, like this:

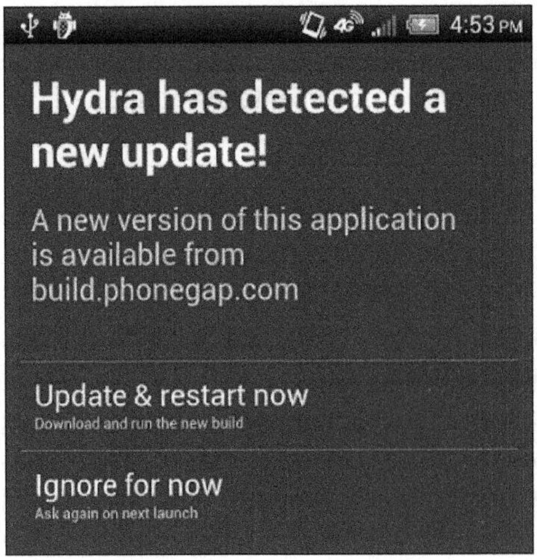

If accepted, the new build will be downloaded and installed directly on the device, as shown here:

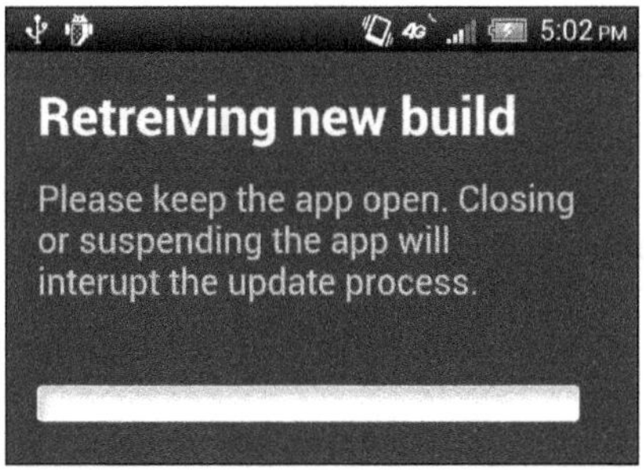

> To find out more about the Hydration service, check out the official documentation, which is available at `https://build.phonegap.com/docs/hydration`.

The PhoneGap Build API

The online PhoneGap Build service is also available via an exposed public **Application Programming Interface** (**API**).

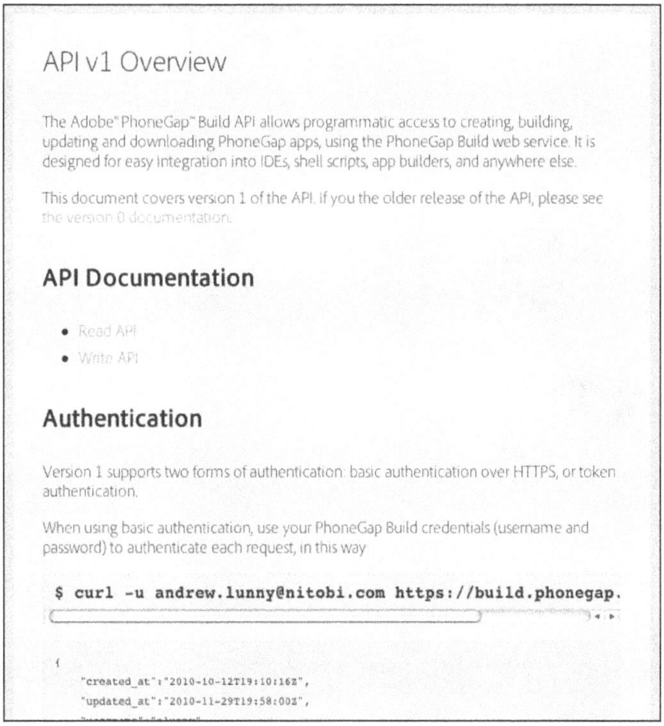

By opening the Build service as an API, developers get the ability to create, build, update, and download PhoneGap applications from shell scripts, the command line, a separate automated Build process, or a source control post commit hook process.

Read and write access to the service is available by using authentication protocols.

> To learn more about the PhoneGap Build API, check out the official documentation at `https://build.phonegap.com/docs/api`.

Index

N

O

P

Q

R

Thank you for buying
PhoneGap 4 Mobile Application Development Cookbook

About Packt Publishing

Packt, pronounced 'packed', published its first book, *Mastering phpMyAdmin for Effective MySQL Management*, in April 2004, and subsequently continued to specialize in publishing highly focused books on specific technologies and solutions.

Our books and publications share the experiences of your fellow IT professionals in adapting and customizing today's systems, applications, and frameworks. Our solution-based books give you the knowledge and power to customize the software and technologies you're using to get the job done. Packt books are more specific and less general than the IT books you have seen in the past. Our unique business model allows us to bring you more focused information, giving you more of what you need to know, and less of what you don't.

Packt is a modern yet unique publishing company that focuses on producing quality, cutting-edge books for communities of developers, administrators, and newbies alike. For more information, please visit our website at `www.packtpub.com`.

About Packt Open Source

In 2010, Packt launched two new brands, Packt Open Source and Packt Enterprise, in order to continue its focus on specialization. This book is part of the Packt open source brand, home to books published on software built around open source licenses, and offering information to anybody from advanced developers to budding web designers. The Open Source brand also runs Packt's open source Royalty Scheme, by which Packt gives a royalty to each open source project about whose software a book is sold.

Writing for Packt

We welcome all inquiries from people who are interested in authoring. Book proposals should be sent to `author@packtpub.com`. If your book idea is still at an early stage and you would like to discuss it first before writing a formal book proposal, then please contact us; one of our commissioning editors will get in touch with you.

We're not just looking for published authors; if you have strong technical skills but no writing experience, our experienced editors can help you develop a writing career, or simply get some additional reward for your expertise.

www.ingramcontent.com/pod-product-compliance
Lightning Source LLC
Chambersburg PA
CBHW062054050326
40690CB00016B/3088